the user's approach to

topological methods in
3d dynamical systems

the user's approach to

topological methods in 3d dynamical systems

mario a natiello
lund university, sweden

hernán g solari
universidad de buenos aires, argentina

World Scientific

NEW JERSEY · LONDON · SINGAPORE · BEIJING · SHANGHAI · HONG KONG · TAIPEI · CHENNAI

Published by

World Scientific Publishing Co. Pte. Ltd.

5 Toh Tuck Link, Singapore 596224

USA office: 27 Warren Street, Suite 401-402, Hackensack, NJ 07601

UK office: 57 Shelton Street, Covent Garden, London WC2H 9HE

British Library Cataloguing-in-Publication Data
A catalogue record for this book is available from the British Library.

ISBN-13 978-981-270-380-4
ISBN-10 981-270-380-2

Printed in Singapore.

Hernán dedicates this effort to his extended family:

- my daughter Flor, who is the essence of life
- Princesa and Leboni (dogs), that taught me that being useful to the pack implies leadership, and that logic applies to dogs, claims of leadership do not make us useful.
- Brillito and Luna (cats), that remind me that freedom cannot be negotiated, and love is something we do not exchange, we just give it away.
- Shoot (horse), who taught me that horse and rider are one and at the same time they are the mirror reflexion of the other.
- and to Bárbara, who made them all exist.

I dedicate this effort to mia amata moglie Patrizia and min älskade dotter Saffo que iluminan cada día de mi vida haciendo que dé gusto vivirlo, and to the forests and lakes of Patagonia and Scandinavia for letting me be part of them every now and then.

Mario

Preface

During most of the Twentieth Century, physicists have been mainly concerned with linear dynamics. Despite the works of Poincaré, Birkhoff and von Neumann, the paradigm in physics was linear dynamics. Courses in Classical Mechanics systematically ignored intrinsically non-linear phenomena and chaos, restricting Mechanics to Integrable Systems, i.e., dynamical systems with an underlying Lie group structure, having dynamics that are exponentials of linear algebras.

During the second half of the 70's the interest in nonlinear dynamics gradually emerged in physics fueled by the possibility of enriching our intuition using increasingly powerful (as well as popular and affordable) computers. The *chaos* paradigm took form, with new problems and new ways to analyze nature. An intense development followed the introduction of graphic workstations in the 1980s. Questions such as: How to characterize systems presenting chaotic dynamics? How to compare models with experiments? were then included within the valid questions of the chaos paradigm.

By that time it became clear that although there exist only a few different ways of displaying linear behaviour (always present in widely different classes of problems), nonlinear problems presented a large variety of different patterns, as well as other specific features such as sensitivity to initial conditions. The urge to generate some comprehensive understanding of chaos (are there different *classes* of chaotic behaviour?) became evident. During the '80s, there were several attempts to solve the classification problem. Earlier attempts focused in the *routes to chaos*, the sequence of bifurcations as a function of a single control parameter, that lead to chaos in a particular system. By the middle '80s this attempt had proven to be of limited use: there were infinitely many routes to chaos in simple two-parameter systems. The chaos community then turned its hopes towards *fractal dimensions*, i.e., a measure of the geometrical imprint (in phase-space) of a chaotic attractor. By the end of the '80s this path had also proven to be almost useless for the characterization/classification problem (although some interesting features such as Barnsley's fractal pictures spun off this effort).

The two main directions taken by the chaos community that we just described were not the only explored directions. Around 1987, a third programme aiming to classify low dimensional (3-D) systems using topological orbit organization began. This project in Physics was preceded by at least two important developments in Mathematics: (a) results from Birman-Williams-Holmes (1983–) developed to extract the knot content of hyperbolic attractors, introducing a geometrical construction that they named template or knot-holder, and (b) results due to Thurston (1979–) on the classification of 2-D diffeomorphisms in terms of two main classes: rotation compatible diffeomorphisms and pseudo-Anosov diffeomorphisms (the latter class admits a fine structure) and the braid content of the diffeomorphism. Thurston's results appeared earlier than the template development, but they were incorporated to the Physics project at a later stage.

While the relation among the mathematical developments and the programme in physics is direct and immediate, there also exist important differences among them. We have given the name **The User's Approach to Topological Methods in 3-D Dynamical Systems** to the classification and recognition programme in Physics, emphasizing that its aim is the use of the mathematical methods (emerging from Topology) in experimental situations. Unlike other programmes in chaos, the topological classification programme is still alive. In this book we intend to re-evaluate this programme.

While writing this book we have come in contact with some difficult aspects concerning how to assess, prove or disprove a certain property in a system, that require a clear conception about how the programme relates to theory, experiment and numerical modeling. The readers will therefore find discussions on, and references to, epistemological matters. We have adopted as much as possible a Popperian *demarcationist* and *fallibilistic* attitude, since we are dealing with experimental science, i.e., our interest is to *induce* from experiments the originating properties of the underlying system in a scientifically valid way. On the contrary, we have left outside this presentation topics that are encompassed by the concept of *normal science* in the sense of Khun (repetition of a paradigm with little variation) as well as some attempts which are still under development, but have not yet reached the level of an organized theory, at least in our understanding. This needs not be a serious loss, there are other sources where the material can be found. It is our hope that all existing proto-attempts will soon reach the mature level so that they can be thoroughly assessed.

Chapter 1 discusses the goals of the programme and why it is needed. Next, we dedicate the first part of the book to a presentation of the mathematical elements that constitute the basis of the programme (Chapters 2–4). Chapters 5 and 6 present the reconstruction problem, proper of the **user's approach**, turning the discussion from mostly mathematical terms to mostly physical (or natural science) terms. Chapter 7 is a guide to some pioneering works in the actual application of the methods to experimental data.

As every programme that actually progresses, there occurs a reformulation process while going from the dreams and illusions of the first days to our present (hopefully more realistic) view. The closing words in Chapter 8 are reserved to a recollection of the conquests achieved by the programme as well as to an evaluation of the problems that the programme faces, having survived 20 years (about twice the survival time of failed theoretical developments, so there is a basis for keeping hopes alive) but having not reached yet a stable status.

Acknowledgments

Along the decades we have worked in this subject we have met a number of colleagues. Many of them became friends along the way, all of them have taught us something that in one way or the other has been important for this book. Thank you all.

The Librarians of **Matematikbiblioteket** as well as the infrastructure at **Lunds Universitets Bibliotek** and at the **Matematikcentrum** of **Lunds Universitet** have been of invaluable support. Thanks.

One of us (MAN), having no extended family in the animal kingdom outside the *homo sapiens sapiens* species, thanks his many friends in different places of the world for their help in making life enjoyable.

MAN gratefully acknowledges travel grants from the Swedish *Vetenskapsrådet*, from *Lunds stads jubileumsfond* and from *Malmö stads jubileumsfond*. HGS thanks the continuous support of the Consejo Nacional de Investigaciones Científicas y Técnicas and grants from the Universidad de Buenos Aires.

Between Villa Elisa and Lund, March 2007.

H. G. Solari, M. A. Natiello

Contents

the user's approach to

topological methods in
3d dynamical systems

Chapter 1

A Crisis in the Experimental Method Archetype

This book deals with the development and use of mathematical methods (in particular, topological) in order to analyze data in the presence of chaotic dynamics.

In this Chapter we will discuss why chaotic dynamics renders the understanding of experimental data extremely difficult, while in the coming Chapters we will develop the methods through which we can (re)construct this understanding.

1.1 The Experimental Method Archetype

At the core of experimental science we find the *reproducibility of experimental results*, i.e., the notion that if the "same" experimental conditions are met, then the "same results" will follow.

For any practical application of this principle, we need to explain what is the meaning of "same results". Normally, we would consider that we have obtained the same results if the values of the observable (measurable) variables in two runs of the experiment do not differ among each other in more than a small amount: the *tolerance* (prescribed in advance). When observable quantities are not constant in time, we expect then the time-trace of the variables to agree.

Words such as "fluctuations" and "random errors" are frequently used in this context, but the main idea behind the archetype is that controlled laboratory conditions allow the researcher (after little or much work) to classify the environmental influence on the experiment in two groups: on one side the relevant group, consisting of a few factors, and on the other all the rest of the universe, regarded as accessory. The experimental response to the relevant group consists in a smooth, distinct signal, while the discrepancies between the observed data and this distinct signal is a small quantity with even smaller or zero time-average.

Reproducibility is closely linked with the idea of *causality*: effects have a cause. Cause and effect are "verified" daily in our lives (or so we believe) and the assumption of natural science is that this always holds: whatever effect we see in the natural world, can be traced back to a cause in the natural world.

Reproducibility helps to identify causes. Apparent violations to reproducibility suggest the experimenter that, maybe, yet another factor is influencing the experiment. This factor may be identified and subsequently incorporated to the relevant group of influences, thus reestablishing the previous status-quo.

The success of this archetype needs no advertisement, just consider how accurately we can predict sun or moon eclipses for centuries or millenia ahead.

Chaos has come to upset this archetype since the sensitivity to initial conditions characteristic of chaotic trajectories warranties that the time traces of two independent runs of the same experiment will develop substantial differences if the dynamics of the system is chaotic. Chaos changes the costs of prediction, making them at last unaffordable, since the precision required for a fixed confidence grows exponentially with the duration in time of the prediction.

In short, the emergence of chaos forces us to find a new meaning for the expression "the same results". We are forced to find less naive forms of comparison for chaotic systems, to find their regularities in spite of the sensitivity to initial conditions and other irregularities.

This Chapter presents the central problem addressed in this book, namely how the emergence of "chaotic" dynamics (meaning with this irregular, deterministic dynamics presenting sensitivity to initial conditions) influences the reproducibility archetype of experimental science.

1.2   Deterministic Chaos

One of the milestones of Dynamical Systems Theory was the rationalization of the idea of *deterministic chaos.*

Definition 1.1   A chaotic set is an invariant, bounded solution set to the initial value problem of a dynamical system, such that this set contains a countably infinite quantity of unstable periodic orbits and at least one dense orbit.

If this set is *attractive* in the sense that initial conditions outside (but nearby) the set get closer to it as the dynamics evolves in time, we may speak of a *strange attractor*[1].

In general, errors in the initial conditions result in errors in the prediction of future behaviour. The new problem posed by systems having strange attractors is that they do not fit in the naive interpretation of the reproducibility archetype.

Small discrepancies in the initial conditions get amplified (because of the "unstable" character of the points in the attractor) and the time evolution gets completely different after a short period of time, although it is almost regular and almost follows a pattern. In technical terms, the exponential divergence of initial conditions is reflected in an exponential loss of accuracy of predictions. Hence, chaos poses enormous or some times impossible demands in the obtainment of accurate predictions. In the worst situations, even with what we may want to consider as "small" errors, the predictions are not good enough already for future times we may want to consider "near" to the present.

Whenever an experimental setup does reveal something like a strange attractor, we are in trouble. Even if we can repeat the experiment with all the "control" parameters accurately fixed, the small deviations in the initial conditions that the rest of the universe produces beyond any possible control will generate a different experimental output. How can we judge the output? How can we rule out experimental mistakes, recording errors, or the like? Or worse, how can we assure the public that this or that is a scientific experiment worth believing in and not just fake?

1.3   Model Validation

The first dramatic consequence of the problems with the reproducibility archetype is the crisis of the usual model validation methods. In fact, starting with Galileo the immediate step after performing a reproducible experiment was to devise a mathematical model where the relevant factors enter in some way in the dynamical description. For example in mechanics, one ends up with Newton's equations of planetary motion, a deterministic

---

[1] Also, the attracting set should be *topologically transitive*, meaning that it cannot be decomposed in smaller "sub"-attractors with no dynamical connection. The existence of dense orbits guarantees transitivity.

model using differential equations. More frequently the models are maybe still deterministic and based on differential equations but just "phenomenological" (describing in some way the observations but lacking an organic theory behind). One way of validating these models is to simulate an experiment with them (via e.g. numerical computations or paper-and-pencil solutions) and compare it with a true experiment performed in similar conditions. This alternative is lost if we do not have tools to compare relatively different experimental outcomes. If the experimental output is "insecure" how can it validate the model output?

Even the occurrence of some well-known pattern such as e.g., the existence of period-doubling cascades in a model, is not enough to validate. Many different, almost identical period-doubling cascades may be hosted within a given system and the question remains: How can we be sufficiently sure that what we "see" is what we believe to see?

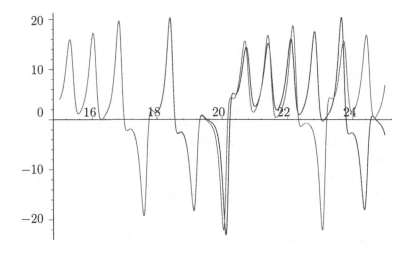

Fig. 1.1   Two different outputs of a numerical integration of the Lorenz equations. The differential equation, parameters and integration methods are the same for both outputs, the only difference being that the initial conditions differ in each graph by 0.001 units (see text).

An even worse difficulty is illustrated in Figure 1.1. Something that distinguishes chaotic problems from non-chaotic ones is the intrinsic impossibility to assign the observed discrepancies to "overlooked relevant factors".

The two graphs in the Figure describe numerical solutions of the Lorenz equations for the set of parameters $\sigma = 10$, $r = 28$, $b = 8/3$. The accuracy of the numerical integration is the same in both runs and we can assume it to be as high as desired (it is a matter of computer time and smart coding to improve this accuracy even more). The differential equation is the same in both cases, the only difference being that the initial condition $(x_0, y_0, z_0)_1 = (0.1, 1.1, 0.8)$ in the first run is modified as $(x_0, y_0, z_0)_2 = (x_0, y_0, z_0)_1 + 0.001(1, 1, 1)$ in the second one.

After about 19 time units the solutions can no longer be considered "equal" and after about 23 time units predictions are completely different: what is large and positive in one run becomes large and negative in the other. This discrepancy is intrinsic to the problem. "Errors" in the initial conditions result in contradicting predictions after a certain period of time and there is no way in which we can come around the problem. Even worse, here there is no "experimental observation" to blame. These are exact (numeric) solutions to a well-posed mathematical problem!

1.4   The Language of Nonlinear Dynamics

Let us review some concepts from the Theory of Nonlinear Dynamics (or Dynamical Systems), assuming a fair knowledge of the subject. The interested reader should refer to books such as [Hale 1969, Guckenheimer and Holmes 1986, Solari et al. 1996a] (or many others) to get a comprehensive treatment of Dynamical Systems Theory.

We assume that recorded experimental data yields time-traces having some deterministic dynamics behind. The experiment records the value of some (function of the) variable(s) $x(t_k)$ at some discrete times t_k. The time-traces will always be discrete, although the underlying dynamics may be described by differential equations for real-valued time (in such a case, the discretization may be refined improving the experimental setup). Hence, our interest will focus on two kinds of systems.

Definition 1.2   (Dynamical System, ODEs) A Dynamical System is a first-order ordinary differential equation

$$\frac{\mathrm{d}x}{\mathrm{d}t} = f(x) \tag{1.1}$$

where t is a real parameter (representing time), x belongs to some manifold M that, for the sake of simplicity, we can assume to be properly described by an open connected set in $\mathbb{R}^n$ or $\mathbb{S}^n$ or some combination of both (some dimensions are described by angles and some other by real intervals) and finally $f : M \mapsto M$ is a (locally) Lipschitz continuous function that will be assumed to be derivable as many times as necessary.

Related concepts are those of *phase space* (the manifold M), *vector field* (the right-hand side of the equation) and *flow* (the set of solutions of the initial value problem with $x(0) = x_0 \in M$), denoted $x(t) = \phi(x_0, t)$. Under these conditions the initial value problem has unique solution.

We will need to consider the action of $\phi(\cdot, t)$ on different point sets of phase space.

Definition 1.3   (Dynamical System, maps) A Dynamical System is a map

$$x_{n+1} = F(x_n) \tag{1.2}$$

where n is an integer parameter (representing discrete time), and $F : M \mapsto M$ is a (Lipschitz continuous) automorphism of phase space M.

Some properties of F that we will encounter frequently (but not always) are that F is orientation preserving and invertible (time-reversible systems have invertible F's).

Maps and ODEs are not completely disjoint worlds. On the contrary, this book deals mainly with the situation where there is a deep connection between both. In fact, certain ODEs can be uniquely associated with a map on a hypersurface in phase space. The intuitive idea is that whenever a solution of an ODE dynamical system closes onto itself in finite time, i.e., it is a *periodic orbit* (see below), then picking a point in that orbit and the velocity vector (a tangent vector to the orbit at that point), we can consider a hypersurface of M having this vector as its normal vector. Initial conditions on that surface lying sufficiently close to the orbit, will return to the surface after a finite time. Poincaré put this idea in mathematical words:

Definition 1.4   (Poincaré section and Poincaré map) Let x belong to a periodic orbit of a ODE dynamical system and let Σ_x be a hypersurface of phase space (or an open connected subset of a hypersurface) containing x and such that:

- For all $y \in \Sigma_x$, $\mathbf{n}_{\Sigma_x}(y) \cdot f(y) \neq 0$, i.e., the outer normal to Σ_x is transverse to the flow (the scalar product has definite sign, say positive, on all of Σ_x).

- Every orbit with initial condition on Σ_x returns infinitely many times to Σ_x for both positive and negative times.
- Every orbit intersects Σ_x.

Then we call Σ_x a Poincaré section for the dynamical system f.

Now consider the map $P : \Sigma_x \mapsto \Sigma_x$ giving the *first return* of a point in Σ_x to the same section. In technical terms, let for every $y \in \Sigma_x$, $T > 0$ denote the smallest positive number such that $\phi(y, T) \in \Sigma_x$. We say then $P(y) = \phi(y, T)$.

Poincaré maps are invertible and orientation preserving as a consequence of the unicity of solutions of the underlying differential equations. However, non-invertible or orientation reversing maps may arise as a consequence of approximations, limiting procedures or reductions of various kinds.

The final intuition is that of *invariant set*, namely a set that is not modified by the dynamics. Such sets may be nice for experiments since if an initial condition lies on an invariant set, then the whole time-trace will remain within the set. Further, the question of *stability* (that we will not discuss) aims to establish whether initial conditions outside but nearby the invariant set will approach the set as $t \to \infty$ or not.

Definition 1.5 (Invariant sets) A set $\mathcal{U} \subset M$ such that $\phi(\mathcal{U}, t) = \mathcal{U}$ for all $t \in \mathbb{R}$ is called *invariant set*. A corresponding definition can be done for maps.

The simplest invariant sets are *fixed points* (sets consisting of one point) and *orbits*. A fixed point x_0 is therefore a zero of the vector field, $f(x_0) = 0$, in ODEs while it has the usual meaning in maps, i.e., $F(x_0) = x_0$. The orbit, $O(x)$, through the point x, is the set $O(x) = \{y \in M$ such that $y = \phi(x, t)$ for some $t \in \mathbb{R}\}$. In particular, the Poincaré map is tied to the idea of *periodic* orbits (where there is a minimum positive T called period, such that $\phi(x, t) = \phi(x, t + T)$). Other remarkable orbits are *homoclinic* orbits. They are orbits having a fixed point x_0 as forward and backward limit in time, i.e., $\lim_{t \to \infty} \phi(x, t) = x_0 = \lim_{t \to -\infty} \phi(x, t)$. Corresponding ideas can be defined for maps.

Orbits are *invariant manifolds* in ODE-systems while they are discrete invariant point-sets in maps (that might belong to some invariant manifold). Finite invariant point sets of the Poincaré map P correspond to periodic orbits of the original flow f. Periodic orbits can hence be labeled with their natural period given by P, namely the return order: A period-k orbit closes in itself after crossing the Poincaré section k times.

1.5   Stereotype Examples of Chaotic Dynamics

Which are the simplest (in terms of their mathematical description) systems presenting chaotic dynamics? 1-D ODE systems can be fully understood with elementary methods. 2-D ODE's have only relatively simple dynamics, as formulated by Poincaré, Bendixson, Peixoto and others [Guckenheimer and Holmes 1986].

Continuing the inventory we come to 1-D maps. In particular, *unimodal maps* [Collet and Eckman 1986] of the interval are continuous maps that present a single maximum, lying in the interior of the interval domain. Further, such maps are monotonically increasing on the left side of the maximum and monotonically decreasing on the right side. Unimodal maps have a lot of structure, including a rigid ordering sequence of its periodic points [Šarkovskii 1964, Metropolis et al. 1973, Li and Yorke 1975, Block et al. 1980] [2], period-doubling cascades [Metropolis et al. 1973, Feigenbaum 1978] and objects that look like "aperiodic" or "infinite-period" orbits. These features arise with very little demands on the map (unimodality and C^k for small k, although already piecewise continuous unimodal maps have many of the standard features).

On one hand, unimodality provides a natural binary partition of phase space: The regions of phase space to the left and to the right of the unimodal maximum can be labeled with two symbols. Orbits of the map (in particular periodic orbits) admit a binary labeling. We speak then of the *itinerary* of the orbit. With a few sensible assumptions, all reasonable itineraries have an associated orbit. Hence, we can describe orbits by their itineraries and describe the action of the map on the space of reasonable itineraries. This procedure is called *symbolic dynamics* and it turns out to be sufficient to achieve a great deal of classification [Collet and Eckman 1986, Solari et al. 1996a].

On the other hand, unimodal maps are two-to-one. Hence, they cannot arise as hypothetical Poincaré maps of some 2-D ODE without further elaboration. Whatever feature of a unimodal map we encounter in a natural process, it must arise in a more complicated context.

---

[2] Šarkovskii's order concerns the *period* of the periodic points, not the orbits themselves. More sophisticated orders have been proposed by Misiurewicz [Misiurewicz 1997] and others.

The next choice is then 3-D ODE's, and the simplest cases are those admitting a Poincaré section. The simplest ODE's presenting complicated dynamics are those admitting a 2-D Poincaré map with a chaotic invariant set. Moreover, such a set should be part of a larger set that attracts a large portion of initial conditions. Two invariant sets caught the attention of most researchers for the last 40 years: Smale's horseshoe and the Lorenz "butterfly".

1.5.1 *Smale's horseshoe*

Chaotic dynamics in bi-dimensional maps is often associated to transversal homoclinic connections. The existence of transversal homoclinic connections under rather general conditions implies the existence of horseshoe maps for subsets of the phase space [Smale 1963, Solari et al. 1996a].

We refer to the general literature in this Chapter for a comprehensive description of Smale's horseshoe construction from the sixties. Basically, a hyperbolic saddle point of a 2-D orientation preserving map, such that the expansion and contraction rates satisfy $\mu < 1/2$, $\lambda > 2$ and where a branch of the stable manifold of the fixed point crosses transversely the unstable manifold, will have a hyperbolic invariant set Λ, called the horseshoe invariant set (this result is called *Smale-Birkhoff's theorem*).

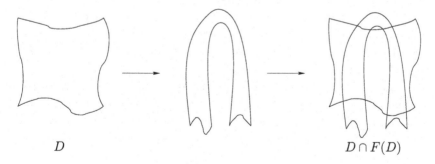

D $\qquad\qquad\qquad\qquad\qquad\qquad D \cap F(D)$

Fig. 1.2   Smale's horseshoe map

The horseshoe set Λ can be seen as the invariant set of a map F mapping the topological unit square D stretched, compressed and bended onto itself as in Figure 1.2. Such a system has periodic orbits of all finite periods and an uncountable number of non-periodic orbits that are *dense* in the set (they come arbitrarily close to all points in Λ).

Labeling the two vertical strips of $S \cap F(S)$, or rather the horizontal strips of $S \cap F^{-1}(S)$ with the symbols 0 and 1, all points in Λ can be described by bi-infinite sequences of 0's and 1's. The symbolic dynamics on Λ is a *shift dynamics*, i.e., the "image by F" of a bi-infinite sequence gives a sequence with the same ordering of the symbols, the only difference being that the reference point of the sequence is shifted one step to the right.

We note on passing that in the limit of "infinite" contraction rate, the horseshoe construction maps the whole unit square onto a unimodal graph. Similarly, looking just to the right half of a horseshoe bi-infinite sequence, we recover a unimodal itinerary.

Let us formalize the idea of "infinite contraction limit" a bit more. An *Anosov diffeomorphism* is a $\mathcal{C}^1$ hyperbolic map on a manifold (for our purposes the unit square or the unit disc would suffice but Anosov diffeomorphisms are easy to construct on the torus) with one contracting direction and one expanding direction. These directions exist on the tangent space of the manifold (technically: tangent bundle), i.e., they vary continuously along the manifold inheriting its differentiable properties. The manifold can be *foliated* with these directions.

A related concept corresponds to Axiom-A diffeomorphisms which are more general. A diffeomorphism is said to be Axiom-A if the non-wandering set is hyperbolic and contains a dense set of periodic points.

A $\mathcal{C}^1$ version of the horseshoe map is an example of an Axiom-A diffeomorphism. The tangent space (bundle) at each point consists of copies of $\mathbb{R}^2$ with the expanding direction along (tangent to) the horseshoe shape and the contracting direction described by the "width" of the horseshoe shape. Infinite contraction limit corresponds to collapsing this width to one point.

A slightly more general class of maps are *pseudo-Anosov diffeomorphisms*[3], where the stable (contracting) and unstable (expanding) foliations have a finite number of points, called *prongs*, where the transversal foliations are singular (see Figure 1.3) Pseudo-Anosov maps may include Axiom-A invariant sets.

---

[3]In many places one finds the expression pseudo-Anosov *homeomorphism* [Hall 1994a] which in fact is more proper, since pseudo-Anosov maps are not differentiable at the singular points. We will describe pseudo-Anosov maps in both ways.

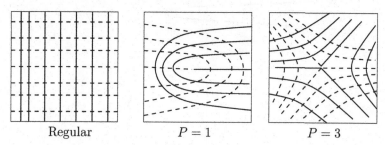

Regular $P = 1$ $P = 3$

Fig. 1.3   Stable (dotted lines) and unstable (solid lines) transversal foliations for regular, and singular interior saddle-points (1-prongs and 3-prongs)

1.5.2   The Lorenz equations

The Lorenz equations [Saltzman 1962, Lorenz 1963, Sparrow 1982] are a simplified model of fluid convection between two plates at different temperatures under the action of gravity. Despite their simple form and the few parameters, their dynamics can be modeled by a singular 2-D return map (see Figure 1.4) and an associated 1-D limit map. The equations read

$$dx/dt = \sigma(y - x)$$
$$dy/dt = rx - y - xz \qquad (1.3)$$
$$dz/dt = xy - bz$$

where σ is a physical parameter (the Prandtl number), b a geometric factor depending on the experimental setup and r relates with another physical parameter (the Rayleigh number). In Lorenz' paper $\sigma = 10$ and $b = 8/3$, while the interesting dynamics arises near $r \sim 28$.

The connection between the equations and both maps finally showing the existence of a chaotic attractor required more than 30 years of research in various fronts [Williams 1977; 1979, Rychlik 1989, Tucker 1999]. The 2-D map can be described as follows. The control section has two halves divided by a singular line (reflecting the fact that the fixed point at the origin has a 2-D stable manifold). The image of each half by the return map is triangle-shaped, partially covering both halves. The invariant set has points in both halves, within both triangles. The dynamics on the Lorenz invariant set can also be described in symbolic terms, using an "alphabet" of only two symbols.

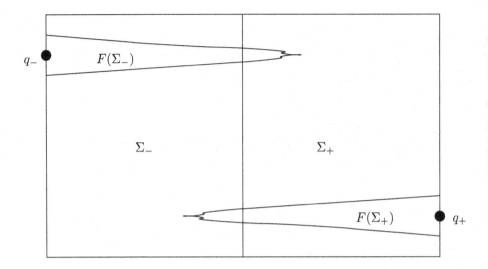

Fig. 1.4   Geometric Lorenz map on a control surface

1.6   Seeking a Way Out / Gathering the Loose Ends

After having realized that chaotic dynamics poses a difficult problem and having inventoried some available material, it is time to propose a plan of action.

What do we want to achieve? We want to be able to *handle* dynamical systems presenting chaos. Not just understanding experimental data (this is already a serious problem) but understanding models and equations that conceal an extremely complicated structure. In particular, we want to find a way out of the reproducibility conflict in chaotic systems.

For dynamical systems that admit a description in $\mathbb{R}^2 \times \mathbb{S}^1$ the very fact that strange attractors contain unstable (more specifically *saddle*) periodic orbits gives us a clue about how to proceed. Time-recordings of such experiments should contain more or less concealed information about those periodic orbits.

Our central goal will then be to *characterize the structure of (invariant sets of) chaotic systems*. In order to do this, we can start from the stereotype problems described above, and try to produce tools to understand systems beyond these problems.

In particular, we will attempt to characterize chaotic systems via time-series, namely a discretized finite portion of a solution, in the cases where this is the only available information. Different finite portions of a dense horseshoe orbit may look completely different. Also, any two orbits on the horseshoe with arbitrarily small differences in their initial condition will eventually separate and lie on opposite sides of the invariant set. If we can identify those orbits as e.g. specific horseshoe orbits, we will recover much understanding despite the individual discrepancies.

These ideas will provide a partial answer to the problems posed earlier in this Chapter as well. Indeed, if we can establish that our widely different experimental (or simulation) data-sets originate in the same chaotic invariant set, then we can recover "reproducibility" on a higher level. Beyond the discrepancies among experimental results, we will have a way to verify (or reject) the idea that the data-sets are coming from the same system. Further, understanding which (classes of) orbits are present in this or that invariant set may help in distinguishing one chaotic problem from another.

1.6.1 *A word of warning*

Knowing what it is to come, it is proper to advance here a word of warning.

The topological methods to be developed in the coming Chapters (Chapter 2 to Chapter 4) rely on a basic assumption, i.e., that we have *secure and unambiguous information about a set of periodic orbits belonging to our system.* This safe information further generates decision tests that allow the researcher to distinguish among fundamentally different properties (or behaviours) and to identify which one suits our original problem.

The data-analysis methods to be discussed in the final part of the book (Chapter 5 and beyond) deal with the complementary problem of connecting experiment with theory. In that part, our goal will be to *generate* secure and unambiguous information about a set of periodic orbits belonging to our system, *given* the available experimental data. This part of the task is much more difficult. It will turn out that different ways of manipulating our data may produce different and *incompatible* sets of information about the periodic orbits.

Does this mean that we build a nice house in the next Chapters only to let it fall apart in the end? Not at all. It simply means that research is hard, and that's why it is fun. There exists no off-the-shelf, black-box method of analysis that from a string of bits generates safe answers about nature without the judicious participation of the researcher. The interpretation we produce is the result of a double game: We will use safe theoretical tools together with data manipulation. The theoretical tools interpret the pair {data + data-analysis} and they will never be able to separate one from the other. It is our responsibility as researchers to understand what we are doing, to specify what we have done to the data and to produce reliable information with clear and specific limits of validity.

Chapter 2

Orbit Organization in $\mathbb{R}^2 \times \mathbb{S}^1$

We ended the previous Chapter stating (roughly) the following working hypothesis: Gathering information about the unstable periodic orbits present in an invariant set (that is "hidden" in a larger attracting set) can help us understand the nature of the underlying system.

This task has many different characteristics that have to be addressed concurrently. On one hand we have dynamical systems issues, coming from the nature of the periodic orbits and of the underlying system. Together with these issues we have data analysis issues, regarding the nature of the basic hypothesis: Does a given set of numbers (integers, in fact) actually represent a periodic orbit of a 3-D dynamical system admitting a Poincaré section? We will address some of the dynamical systems issues first, leaving the data analysis to a later Chapter, since this is a book on dynamical systems that *uses* data handling rather than the other way around. The analysis process will raise new issues that we will consider as they show up. So we assume in this Chapter that we have a 3-D dynamical system of which we know a finite number of hyperbolic periodic orbits.

Many definitions in this and the following Chapters will be given discursively, i.e., without setting up an explicit *definition* environment. We will instead write the new concept in *italics*, subsequently defining the ideas behind the concept.

2.1 Examples of Dynamical Systems in $\mathbb{R}^2 \times \mathbb{S}^1$

Before developing the analysis tools, it is mandatory to present some examples from applications that display the relevance of focusing on systems in $\mathbb{R}^2 \times \mathbb{S}^1$. Apart from Lorenz equations, that may be considered as an "academic" simplification of the dynamics involved in the Bénard experiment from fluid dynamics, there are a number of modeling situations where a $\mathbb{R}^2 \times \mathbb{S}^1$ description seems proper.

2.1.1 *Periodically forced nonlinear oscillators*

A one degree-of-freedom nonlinear mechanical oscillator subject to an external periodic force may be described using Newton's equations in the following way:

$$m\frac{d^2x}{dt^2} = -kx - \beta\frac{dx}{dt} - \gamma g(x) + A\cos(\omega t) \tag{2.1}$$

where x is the deviation of the oscillator from its equilibrium position, m is the mass, k is the oscillator constant, $\beta > 0$ models the damping effects, $\gamma g(x)$ describes the departure of the oscillator from linear behaviour (usually $g(x) = x^3 + O(4)$) and A is the amplitude of the external force. ω may be rescaled to unity changing the time-scale.

The canonical example of this kind of oscillators is the Van der Pol oscillator, thoroughly described in e.g., [Guckenheimer and Holmes 1986]. Apart from mechanics, oscillators of this kind are useful in describing nonlinear electrical circuits, among other things. Rewriting the above equation as an adequately rescaled autonomous dynamical system, we have:

$$\begin{aligned} \frac{dx}{dt} &= y \\ \frac{dy}{dt} &= -ax - by - cg(x) + A\cos\theta \\ \frac{d\theta}{dt} &= 1 \end{aligned} \tag{2.2}$$

For any fixed choice of the angular variable θ_0, the xy-plane is good as a Poincaré section and the dynamics can be described either using the full three-dimensional flow or with the 2-Dimensional Poincaré map on the chosen control section.

2.1.2 *Laser with modulated losses*

Laser physics has been a source of problems in low dimensional chaos [Arecchi et al. 1982; 1986, Arecchi 1988, Oppo et al. 1986, Tredicce et al. 1986]. The description of a single frequency laser can be performed with a few basic, phenomenological, variables: electric field, E, atomic polarization, P, and population inversion, N [Baldwin 1969, Risken 1989]. In some lasers the response time of these variables allows for further dimensional reduction (called adiabatic elimination in laser physics [Oppo and Politi 1989], a particular case of reduction to the center manifold). Class B lasers, such as the CO_2-laser can be efficiently described in terms of the light intensity, $I = |E|^2$, and the population inversion, N by the following set of equations [Solari et al. 1996a]

$$\frac{dI}{dt} = I(\epsilon + \beta N)$$
$$\frac{dN}{dt} = \gamma(N_0 - N) - \beta N I \qquad (2.3)$$

In the experimental setup of the laser with modulated losses, the reflectivity of the laser's cavity is periodically changed using an electro-optic modulator, hence $\epsilon = \epsilon_0 + \epsilon_1 \cos(\omega t)$ [Solari et al. 1987] producing a 3-D autonomous system described in $\mathbb{R}^2 \times \mathbb{S}^1$.

2.2 Homotopies and Topological Properties

To begin with, the actual shape of the orbits should not be important. Such things as shapes can be altered by coordinate transformations and in fact coordinate choice is little more than a tool of the researcher. The identification of relevant dynamical properties should occur in a higher level than that of coordinate choice. So we look for orbit properties that persist upon (valid) coordinate transformations,

More specifically, we look for properties that persist under homotopies. A *homotopy* is a continuous transformation that preserves *some* property that is specified in each particular case. For example a circle imbedded in the plane is homotopic to a square since there exists a map $F : [0,1] \times \mathbb{R}^2 \mapsto \mathbb{R}^2$ such that $F(0,x)$ is the identity in $\mathbb{R}^2$, $F(1,C)$ is a square (where C denotes the original circle) and $F(t,C)$ is e.g. a Jordan curve for all $t \in (0,1)$ (what is preserved here is the property of being a closed plane curve without self-intersections). We will encounter many versions of the homotopy property along the book.

Homotopies are useful also in what regards to the assumption of hyperbolicity. Consider a dynamical system that depends on certain parameters, having a hyperbolic periodic orbit. If the parameter values are away from bifurcation points, dynamical systems having nearby parameter values will also have a hyperbolic periodic orbit. Moreover, as a consequence of the

Hartman-Grobman theorem [Guckenheimer and Holmes 1986] these orbits can be deformed into each other by a coordinate transformation. In other words, they are homotopic.

Homotopy defines an equivalence relation. What we are looking for are properties of the equivalence class to which a given orbit belongs, rather than properties of the orbit itself. In this way, we get rid of spurious anomalies that could arise from (a) the choice of coordinates, (b) imprecision in the determination of system parameters.

2.3   Periodic Orbits as Knots

A periodic solution of a 3-D dynamical system is a paradigmatic example of an *oriented knot* [Holmes and Williams 1985, Kauffman 1991, Ghrist et al. 1997, Adams 2001, Carlson 2001]. A knot is in fact an abstraction originated in the idea of a continuous closed curve in $\mathbb{R}^3$ without self-intersections. The time-parameterization defines a circulation along the orbit (the *orientation*), with which knot-information can be retrieved. The question is which information is relevant for our purposes.

Suppose we consider a periodic orbit O and study it by moving in 3-D space along the orbit, starting at the point $x_0 \in O$ for $t = 0$. In addition to the discussion in the previous Section, we have the freedom of choosing x_0 and/or the origin of time. The information we seek has to be independent of this choice as well.

If we project the orbit in 2 dimensions (e.g. on a plane), we obtain a closed curve with self-intersections (except for the trivial case of a topological circle). Self-intersections can be assumed to occur pairwise (at most two arcs corresponding to different portions of the orbit cross at a given point), since if this is not the case, a coordinate transformation can split multi-crossings into a set of pairwise crossings. In knot theory this curve is called a *knot projection* [Adams 2001]. The self-intersections can be displayed in their order of appearance and labeled with + or − indicating which strand lies above the other. For example, left-over-right in the direction given by the orientation takes "+" and right-over-left takes "−". We may then speak of a *signed crossing*. In this way, we obtain a description of the knot. For example, consider the left string of Figure 2.1, I. Taking the orientation to be *top* → *bottom* the crossing gets the label "+".

Some of these intersections may be artifacts due to our choice of coordinates (or of projection), which might disappear with a different choice. In

knot theory we say that there are different knot projections corresponding
to the same knot. The *Reidemeister moves* [Adams 2001] shown in Figure 2.1 are changes in the knot projection that do not alter the knot. For
example, the second Reidemeister move corresponds to an apparent self-
intersection that may be removed via a suitable change of coordinates in
$\mathbb{R}^3$. The remaining Reidemeister moves are (I) tightening up a loop in an
arc and (III) sliding an arc from left to right on top of a crossing of two
other arcs. These moves represent the action of homotopies of the orbit on
the knot projection.

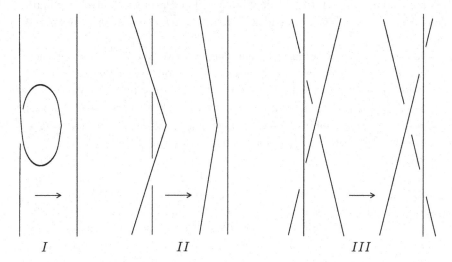

Fig. 2.1   The three Reidemeister moves. The second move corresponds to an apparent
self-intersection of portions of an orbit.

As a matter of fact, to realize that the Reidemeister moves produce knots
that are homotopic to each other is slightly more than a nice exercise. The
deeper insight that the moves convey lies in the fact that finite combinations
of these three moves *exhaust* all possible actions that preserve the homotopy
class.

We can associate to a given periodic orbit an equivalence class of knot
projections. All projections that differ in any number of Reidemeister moves
are equivalent. Moreover, list all the crossings in the projection and consider
all cyclically shifted lists: Instead of $1, 2, 3 \cdots, M$ take e.g. $p + 1, p +
2, \cdots, M - 1, M, 1, 2, \cdots, p - 1, p$, for $p = 1, \cdots, M - 1$. All these lists
correspond to the same equivalence class. This takes care of the choice
of x_0 mentioned above. Technically, given a periodic orbit we can define a

knot [Carlson 2001] as the equivalence class associated to homotopies of the orbit (the more technical concept of *isotopies* and even *ambient isotopies* can be used in the definition).

As the reader may notice, given two periodic orbits, or given two knot projections with a large number of crossings (as of 2001 "large" could mean 17 since all non-factorizable knots up to 16 crossings have been tabulated [Adams 2001]) it might be a formidable task to decide whether they are in the same equivalence class or not. We need some object that can be computed for each knot, that gives different output for different equivalence classes and is simpler to handle than the knot projection itself with its signed crossings. Hence, the idea of *knot invariant* took form along the 20th century.

For example, the linking number between two knots can be obtained adding up the signed crosses between both knots and dividing by two, since each cross adds or subtract a π-turn of one knot over the other.

The Conway polynomial can be seen as a bookkeeping of the operations necessary to de-assembling a knot. Given a knot, say L_+, presenting a positive crossing, two new knots are constructed, the first one changing the positive crossing to negative (creating L_-), and the second one is obtained by "smoothing out" the crossing (see Figure 2.2), producing L_0. To each knot or link a polynomial, $\nabla(z)$, is assigned, where the polynomial 1 corresponds to the unknot and the relations $\nabla_{L_+}(z) - \nabla_{L_-}(z) = -z\nabla L_0(z)$ are satisfied. The recursive relation allows to compute the polynomial of the desired knot de-assembling the projection [Carlson 2001]. The process is illustrated in Figure 2.3.

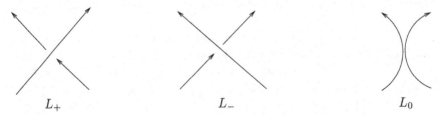

Fig. 2.2   The links produced by changing a positive crossing (left) into a negative crossing (center) and smoothing the cross (right)

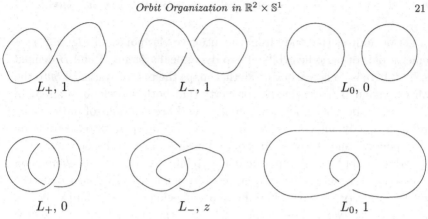

$L_+, 1$ $L_-, 1$ $L_0, 0$

$L_+, 0$ L_-, z $L_0, 1$

Fig. 2.3   Conway polynomials.

A number of knot invariants have been described and used with relative success [Adams 2001], but all knot invariants so far have the same drawback: *there exist inequivalent knots having the same knot invariant.* Hence, regarding the analysis of individual periodic orbits, the most we can hope for in terms of knots is that *if two given orbits have different knot invariants, then they are not homotopic.*

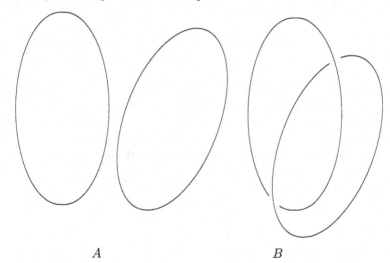

A   *B*

Fig. 2.4   A pair of periodic orbits for dynamical systems A and B.

We can go a bit further using just knots. If we have a set of (more than one) periodic orbits, we may want to study them as a set rather than just individually. A non-empty set of knots is called a *link*.

Let us address the study from the intuitive viewpoint. In Figure 2.4 we depict a pair of periodic orbits of two dynamical systems A and B defined in $\mathbb{R}^3$. There is no homotopy in $\mathbb{R}^3$ that maps the pair of system A onto the pair of system B preserving the property that both periodic orbits remain as such (i.e., do not break apart in arcs) and are solutions of initial value problems of a dynamical system. In fact, to map one pair onto the other by e.g. "moving" one of the orbits from its position in B to its position in A, this orbit would be forced to have non-zero intersection with its companion orbit somewhere along the way. A non-zero intersection would violate the unicity property: A given initial condition belongs to one and only one orbit; never to two different orbits. The fact that the orbits are linked or not linked is a link invariant that helps to decide whether a given pair of orbits can occur in a dynamical system or not. The technical concept is the *linking number* [Guckenheimer and Holmes 1986, Solari et al. 1996a] between two orbits.

Definition 2.1   (Linking number) Given two periodic orbits O_1 and O_2 in $\mathbb{R}^3$ and an orientation, the linking number is the integral:

$$L(O_1, O_2) = -\frac{1}{4\pi} \int_{O_1} \int_{O_2} \frac{(\mathbf{x}_1 - \mathbf{x}_2) \cdot (d\mathbf{x}_1 \times d\mathbf{x}_2)}{|\mathbf{x}_1 - \mathbf{x}_2|^3} \qquad (2.4)$$

where $\mathbf{x}_1$ and $\mathbf{x}_2$ run along each respective orbit according to the given orientation, while $\cdot$ and $\times$ indicate the usual scalar and vector products.

It might be more or less easier to realize it, but the integral yields just an integer. It counts the number of revolutions that one orbit does around the other after a complete circulation. The idea goes back to Gauss when computing the effect of magnetic fields in coils. A nicer result [Adams 2001] is that the linking number can be computed using the signed crossings of a knot projection simply as:

$$L(O_1, O_2) = \frac{1}{2} \sum_{i=1}^{N} \sigma_i \qquad (2.5)$$

where we assume there are N signed crossings, and σ_i is 1 or -1 according to the sign assignation.

The reader may want to prove that the number of crossings of a pair of orbits is always even.

Further information about knots and links can be retrieved from the home-page of various active researchers in the field, for example, Morwen

Thistelwhite: http://www.math.utk.edu/~morwen/ [1].

2.4 Periodic Orbits as Braids

The additional information given by the existence of a Poincaré section allows us to lift the analysis beyond knots and consider a more structured mathematical object: The braid.

In the situation we are studying, time-evolution can be identified with a monotonically varying angle θ. The Poincaré section is given by a fixed angle θ_0 along $\mathbb{S}^1$. Having this global Poincaré section generates a number of restrictions. First, all knots have the same orientation. Secondly, we can associate to each knot an integer number, the *period,* indicating how many times the knot crosses the control section (a 2π-revolution in θ). Finally, the knot projection acquires a very illustrative shape. Let us represent $\mathbb{S}^1$ by a vertical interval with the endpoints identified. Further, let the projection of $\mathbb{R}^2$ be a horizontal line segment (of which there are two identical copies, one at θ_0, another at $\theta_0 + 2\pi$). A period-n knot will cross the (projected) control section (both copies) in n different points. The knot can then be represented by n strands joining the upper n points with the lower n points (without self-intersections). Each lower point is connected to only one upper point by a strand. The crossings in the knot projection are described by similar crossings among the n strands. See the example in Figure 2.5.

Definition 2.2 Braid: A braid of n strands is a homotopy class of continuous functions $B_n : [0,1] \mapsto (\mathbb{R}^{2n} \backslash \Delta)$, where Δ is the great diagonal of $\mathbb{R}^{2n}$ i.e., $x \in \Delta \Leftrightarrow x = (x_1, x_2, \cdots, x_n), x_i \in \mathbb{R}^2, i = 1, \cdots, n, (\exists i \neq j) : (x_i = x_j)$.

One may for simplicity set $B_n(0) = B_n(1)$ and further choose these image points to lie along a straight line. Each component of the image defines a strand of the braid. The role of Δ is to assure that there are no intersections among the different strands. To recover a periodic orbit from the braid, one just connects each point in $B_n(1)$ with the corresponding point in $B_n(0)$ as indicated in Figure 2.5. The braids of n strands form a group, where the group operation is juxtaposition of two braids, one after the other.

---

[1]Of course, home-pages are rather ephemeral. People move, change job, retire, etc., and their home-pages may not always persist as long as their books do.

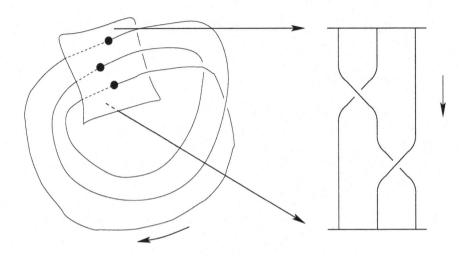

Fig. 2.5 A period-3 orbit of a dynamical system in $\mathbb{R}^2 \times \mathbb{S}^1$ and its knot projection as a braid. The arrows along the flow and braid indicate the time evolution from θ_0 to $\theta_0 + 2\pi$. These two angles are split in the braid graph (right, bottom and top) but identified in the time evolution (left, control section).

2.4.1 *Braid Words*

We can label the braid crossings in a similar way as in knots, this time using the additional strand information. We call σ_i the crossing where strand i goes *over* strand $i + 1$, and σ_i^{-1} the opposite crossing where i goes *under* $i + 1$. For a braid of $n > 1$ strands i runs from 1 to $n - 1$. Hence, a braid can be described by listing the crossings of its knot projection in the parametric order. A braid can then be represented by a *braid word*. The braid in Figure 2.5 has the word $W = \sigma_1 \sigma_2^{-1}$. We say that a *positive braid* has no negative exponents among the σ's in its braid word.

2.4.2 *The braid group I*

Braids are also introduced as the free group of n generators with the following two restrictions: $\sigma_i \sigma_j = \sigma_j \sigma_i$, $|i - j| > 1$ and $\sigma_i \sigma_{i+1} \sigma_i = \sigma_{i+1} \sigma_i \sigma_{i+1}$. The latter restriction resembles Reidemeister move III [2] in the sense that it states that a crossing between two strands can be moved to the "other side" of a third strand simply by sliding it down, as if the strands were

---

[2]Move I is unnecessary since defining braids as functions rules out the possibility of self-loops and move II is immediate since the group property assures that $\sigma_i \sigma_i^{-1} = Identity$.

real pieces of rope. The other restriction is self-evident in terms of strings and homotopies but requires explicitation when viewing the braid group in abstract algebraic form. See Figure 2.6 for an illustration.

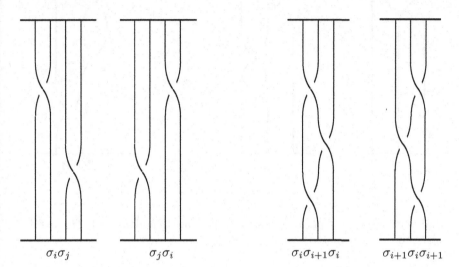

$$\sigma_i \sigma_j \qquad\qquad \sigma_j \sigma_i \qquad\qquad\qquad \sigma_i \sigma_{i+1} \sigma_i \qquad \sigma_{i+1} \sigma_i \sigma_{i+1}$$

Fig. 2.6   The fundamental relations of the braid group.

2.4.3 The braid group II

Braids can be equipped with a group structure in a more intuitive way. Consider two braids of n-strands, a and b. In graphical terms, the (right) product ab consists in placing a on top of b erasing the intermediate section (see Figure 2.7) to produce the braid ab. Clearly, the identity corresponds to a braid with no crossings while the inverse element is easier to write in terms of the free group generators, σ_i. The braid $a = \sigma_{i_1} \sigma_{i_2} \dots \sigma_{i_K}$ has the inverse $a^{-1} = \sigma_{i_K}^{-1} \dots \sigma_{i_2}^{-1} \sigma_{i_1}^{-1}$, i.e., a braid with all the crossings reverted and placed in reversed order.

2.5   Coloured Braids, Linking Numbers and Relative Rotation Rates

Consider the braids with n strands as a group, B_n. This group has a proper subgroup P_n, i.e., $bP_nb^{-1} = P_n$ for all $b \in B_n$. The elements in P_n are the

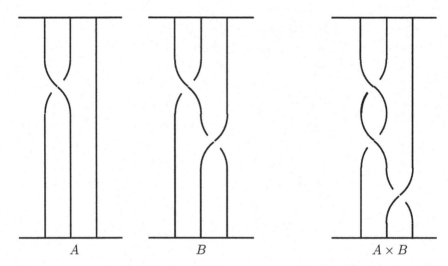

$$A \qquad\qquad B \qquad\qquad A \times B$$

Fig. 2.7   The product of two braids *ab* for the braid group.

braids with associated permutation Id (the identity) and are called *coloured braids* [Fathi and Shub 1979]. The quotient group $P(n) = B_n/P_n$ is the permutation group $P(n)$.

A group having a normal proper subgroup can be represented by an extension of the quotient group by the subgroup in the form of pairs (p, l) where p belongs to the normal subgroup and l to the quotient group. The general procedure is described in [Kirillov 1976]; here we are interested in representations of the braid group obtained as extensions of the permutation group. Choosing a "braid representative" of each permutation $p \in P(n)$ we can associate to it a coset pP_n of B_n by multiplying (in B_n) the braid representative of p with each coloured braid in P_n. This procedure exhausts the whole group B_n. Similarly, from a given (colourless) braid b in B_n one can obtain a corresponding (coloured) braid l in P_n by multiplication (in B_n) with p^{-1}, i.e. $l = p^{-1}b$, where p is the global permutation of the n strands produced by b. A representation based on right multiplication and right cosets can be produced in a similar way. Hence, the braid group B_n can be fully decomposed/reconstructed in terms of the above mentioned pairs (p, l).

2.5.1   *Matrix representation of braids*

With every braid we associate two matrices p and c. p is the matrix giving the permutation of the braid while c is a symmetric crossing matrix with its upper triangle defined as follows: c_{ij} is the sum of exponents for the σ's involved in the crossings between strand i and strand j.

Each braid representative will be of the form (p, c). The identity is represented by $(Id, 0)$. This representation contains only pairwise information on strand-crossing and cannot be faithful. For example, the non-trivial braid of 3 strands $(\sigma_1^{-1}\sigma_2)^3$ has the same matrices as the identity braid.

The group composition law for braids in this representation can be read directly from its construction as:

$$(p', c') \times (p, c) = (p * p', c + p^T * c' * p). \tag{2.6}$$

The primary interest on this representation is that it has an associated class-invariant (but not knot-invariant). We call this invariant the *intertwining matrix* C_n defined as ($\times$ is here the group product):

$$(\frac{1}{n})(p, c)^{\times n} = (Id, C_n). \tag{2.7}$$

If the braid represents just one periodic orbit, n is the period (or equivalently the number of strands).

2.5.2 *Relative rotation rates*

Relative rotation rates were first defined for periodically forced bi-dimensional systems [Solari and Gilmore 1988b]. These systems can be recasted as autonomous systems with a third coordinate $\phi = \omega t$ mod 2π.

Poincaré sections can be obtained as stroboscopic sections, $\phi = \phi_0$. A period n orbit presents n different intersections with the Poincaré surface. Let A, B be periodic orbits, pertaining to the same system, of period n_A and n_B respectively. And let a_i and b_j be the respective intersections with the Poincaré plane.

The integral

$$R_{ij}^{AB} = \frac{1}{2\pi n_A n_B} \int_0^{n_A n_B 2\pi/\omega} \frac{\mathbf{r}_{ab} \times \frac{d\mathbf{r}_{ab}}{dt}}{|\mathbf{r}_{ab}|^2} \, dt \tag{2.8}$$

with $\mathbf{r}_{ab} = \mathbf{x}_a(t) - \mathbf{x}_b(t)$ the vector from a point in the orbit B to a point in the orbit A at time t, and $\mathbf{x}_a(0) = a_i$, $\mathbf{x}_b(0) = b_j$, defines an integer number for each pair of initials conditions taken. R_{ij}^{AB} is called the *relative rotation rate* and it corresponds to the average number of turns given by the vector $\mathbf{r}(t)$ upon return to its initial value after a time of $n_A n_B 2\pi/\omega$.

These numbers are well defined since $\mathbf{r} \neq \mathbf{0}$ as a consequence of the unicity of the solutions of differential equations.

An alternative form of counting the relative rotation rates is as follows. Display the two orbits from $t = 0$ to $t = n_A n_B 2\pi/\omega$ projected in the first coordinate and the phase and keep track of the sign corresponding to the difference between the second coordinates at each crossing. The relative rotation results then of counting the signed crosses.

The latter construction does not require a bi-dimensional periodically forced flow and is generalized to three dimensional flows with global Poincaré sections.

The intertwining matrix defined above is the collection of *relative rotation rates*. It can be easily verified that it is a class-invariant (up to permutation) and as such, it partially characterizes the organization of periodic orbits. Equation (2.7) provides a useful algorithm for computing relative rotation rates.

Following a pair of periodic orbits A and B of periods p and q during $mcm(p, q)$ periods (where mcm is the minimum common multiple) one will return to the original relative situation of the two orbits. This evolution can thus be associated to a coloured braid.

Lemma 2.1   *The linking number between two periodic orbits A and B can be expressed as one half of the sum of the relative rotation rates between the strands with initial point in A and the strands with initial point in B [Solari and Gilmore 1988b;a].*

The proof of this Lemma runs just by showing the equality. Let s_{ij} be the sum of exponents corresponding to crossings between the strands beginning at a_i and b_j. Then,

$$R_{ij}^{AB} = \sum_{m=1\ldots n_A n_B} \frac{s_{i+mj+m}}{n_A n_B}. \tag{2.9}$$

We have that

$$\sum_{i,j} R_{ij}^{AB} = \sum_{i,j} \sum_{m=1\ldots n_A n_A} \frac{s_{i+mj+m}}{n_A n_B}. \tag{2.10}$$

Rearranging indices we obtain

$$\sum_{i,j} R_{ij}^{AB} = \sum_{m=1\ldots n_A n_B} \sum_{i,j} \frac{s_{ij}}{n_A n_B} = \sum_{i,j} s_{ij} = 2L_{AB}, \tag{2.11}$$

where the latter equality corresponds to a well-known property of the linking number [Carlson 2001].

In other words, take the braid associated to the union of both orbits (having $m = p + q$ strands). Use Eq. (2.7) with $n = mcm(p, q)$ and sum all matrix elements of the upper diagonal triangle of C_n finally dividing by two since each crossing adds a π relative rotation.

Equivalently, the "flow"-method to compute linking numbers [Solari and Gilmore 1988b;a] can be rephrased as follows: Take one point in orbit A and one in orbit B. Consider the braid of 2 strands obtained by following their time-evolution over n periods. The associated c matrix in the above representation is now a 2×2 matrix. The linking number is $(c)_{12}/2$.

2.6   The Knot Holder

Birman and Williams realized that there was a one-to-one correspondence between the periodic orbits in flows in $\mathbb{R}^2 \times \mathbb{S}^1$ having a contracting direction and the orbits in a branched manifold that can be thought of as the "limit for infinite contracting rate" of the flow [Birman and Williams 1983a;b]. Such a branched manifold is called a *template* or *knot holder*.

Let $\phi_t : M^3 \to M^3$ be a flow on a 3-manifold such as $\mathbb{R}^2 \times \mathbb{S}^1$ having a hyperbolic invariant set with a neighbourhood $N \in M$. Let $\sim$ denote the equivalence relation $z_1 \sim z_2$ if $|\phi_t(z_1) - \phi_t(z_2)| \to 0$ as $t \to \infty$, and $\phi_t(z_i) \in N$ for all $t \geq 0$. Effectively, this equivalence relation induces a collapse of the flow along the stable manifold, and identifies orbits with identical future. The flow becomes a semi-flow on a two-dimensional manifold. What is remarkable about this tremendous collapse is that *the periodic solutions within the invariant set will not change their topological properties under the projection.*

The reason is the following. Let x be a point on a periodic orbit, and $W^s = \{y : d(\phi_t(y), \phi_t(x)) \to 0 \text{ as } t \to \infty\}$. This set corresponds to the ω-limit of the periodic orbit, i.e., those points whose dynamical evolution approach the periodic orbit as $t \to \infty$. In simple words it is like the "stable manifold" of the orbit. Clearly, $W^s(x)$ will not intersect any other periodic orbit, or its corresponding stable manifold, as two hyperbolic periodic orbits are separated in phase space and no point in phase space can have two different "futures". Therefore, the map ϕ_t is 1-to-1 within the set of periodic orbits of the flow and moreover the collapse preserves each periodic orbit "as it is" and it does not change the linking or knot type.

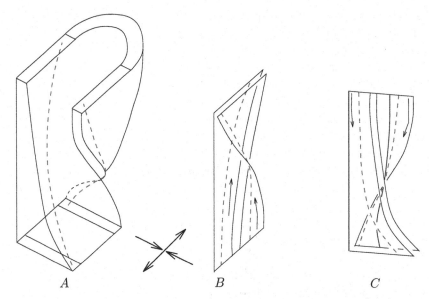

Fig. 2.8   A model of a flow hosting a Horseshoe (**A**), the branched manifold representing the invariant set (**B**) and the usual pictorial representation of templates, with time flowing "downwards" (**C**). The thick lines represent period-1 orbits while the dotted line represents a period-2 orbit.

Figure 2.8 illustrates the discussion. In (**A**) we show the simplest flow compatible with a horseshoe map. The branched manifold obtained through the collapse along the stable manifold is displayed in (**B**). In both figures, time flows "upward". The manifold in (**B**) is recast in (**C**) with time flowing downwards, as it is more frequent to find it in this way in the literature. The illustration includes a representation of period-1 and period-2 orbits.

2.6.1   *Applications*

The usefulness of the template construction lies in the fact that the branched manifold is a simpler representation of the underlying dynamical system through which much of the braid information about the periodic orbits of the system, including linking numbers and relative rotation rates can be obtained just by looking at a graph.

In particular, if one has to establish which one of two different templates represents a given system, certain periodic orbits render the decision evident. A given periodic orbit may not be present in one template but present in the other, or, even if the orbit is present in both templates, its linking with other orbits could be different in each template. In simpler words, knowing the template you can say a lot about the periodic orbits of the system.

For example, consider the period-1 and period-2 orbits of the geometric Lorenz attractor [Birman and Williams 1983a;b], and the corresponding orbits for the Horseshoe system. In the latter system one of the period-1 orbits is linked to the period-2, while in the Lorenz system all three orbits are unlinked, see Figure 2.9. The strength of the Knot-Holder approach is that these matters can be addressed graphically, just by drawing the corresponding orbits on the template.

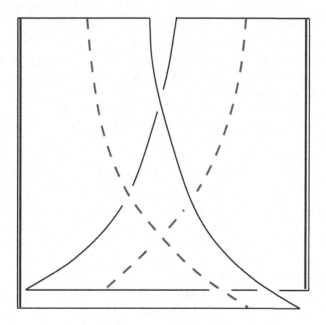

Fig. 2.9   Templates and low-period orbits for the Lorenz system.

The natural extension of these ideas from the applications point of view is to record the templates of standard problems and subsequently to analyze the relative linking properties of periodic orbits given by each template. In this way, given a finite set of periodic orbits taken e.g. from experimental data (*how* this is done is another stuff that will be addressed in a future Chapter) one may discard this or that template simply by checking its linking properties against those of the data. The most one can get in this way is a notion of compatibility:

Definition 2.3   A flow is compatible with a branched manifold when all the periodic orbits of the flow can be associated with periodic orbits in the branched manifold. Two flows will be equivalent if they are compatible with the same branched manifold [Mindlin et al. 1991].

In practice, two time-series taken from the same dynamical system may display different (even disjoint) finite sets periodic orbits. However, both sets may be sufficient to identify the same underlying template. On the other hand, time series from e.g., the Horseshoe and the Lorenz systems containing just low-period orbits (at least period-2) will be enough to discriminate the data.

Still, we need more powerful tools of analysis. Time-series, being finite, might be compatible with different (incompatible) templates. We will analyze this question further in the next Chapter.

2.7   Appendix: The Horseshoe Template and Orbit Classification

We discuss here as an example the orbit classification for the horseshoe map. The template was already depicted in Figure 2.8 along with some low-period orbits. Graphically (use (**C**) in that figure to fix ideas) one may represent the template as a rectangle that is ripped from top to bottom, both teared halves are stretched at the bottom so that they cover the whole rectangle, and the right half is twisted clockwise by an angle of π.

All horseshoe orbits can be described by their bi-infinite itinerary using two symbols (referring to the orientation preserving and orientation reversing strips of the horseshoe) [Guckenheimer and Holmes 1986, Solari et al. 1996a]. Periodic orbits of period k correspond to periodic itineraries and they can be labeled by a finite string of 0's and 1's containing k elements. A given orbit can be labeled by at most k different strings (depending on which point on the orbit we choose to be the starting one). Standard choices are to label the orbit with the symbolic sequence of its leftmost point (hence, orbits start with 0 except the period one 1), adopted here, or using the rightmost point (hence, orbits start with 1 [Hall 1994a]).

Elaborated results such as the fact that all horseshoe braids are positive are immediate by considering the associated template. Indeed, one may

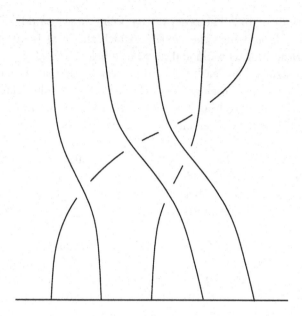

Fig. 2.10   The period-5 orbit 00101 of the Horseshoe template and its corresponding braid.

realize graphically that all strand-crossings on this template will occur "left over right" because (a) the twisted branch lies behind the non-twisted one and (b) the twist is clockwise. The apparently innocent modification of twisting the second branch counterclockwise gives already a much more complicated structure than the horseshoe template, admitting both positive and non-positive braids, since by reversing (b) strands lying on the twisted branch will cross "right over left"[3]. To compute the braid and linking properties of a horseshoe orbit given its name, we proceed as follows:

(1) The k elements correspond to k points on the top and bottom lines of the template.
(2) On the top line, zeroes correspond to points on the orientation preserving branch of the template and ones to the orientation reversing branch (both branches are identified on the bottom line). Points are laid in ascending order (left to right) on each branch.
(3) Cyclic permutation of the k elements gives the different points involved

---

[3]A comparatively simple template with only three branches suffice to host all possible knots and links [Ghrist and Young 1998]. Note however that this does not mean that such template will hold all possible braids (see Chapter 6).

in the periodic orbit. The order of the points in the braid (i.e., which point is the first, which the next,... which the last) is given by the itinerary order in 1-D unimodal maps (see below).

(4) The braid strings joins each point to its consecutive point in the orbit (cyclically) and the crossings (σ or σ^{-1}) are given by the relative position of both template branches (orientation preserving branch always above the orientation reversing).

The itinerary order[4] [Metropolis et al. 1973, Collet and Eckman 1986, Solari et al. 1996a] of a string of 0's and 1's is as follows. Let p be a (possibly empty) string and let the number of 1's in p be c_p. If c_p is even (including zero) then $p0 < p1$, if c_p is odd, then $p1 < p0$. Also if $p > q$ then $pX > qX$ for any string X.

For example, the orbit 00101 has period-5. Labeling the points from left to right with the numbers 1 to 5, points $1, 2, 3$ lie on the orientation preserving sheet and points 4 and 5 are on the orientation reversing one. The points of the orbit are 00101, 01010, 10100, 01001 and 10010. Using the unimodal order, the permutation of the orbit is: $1 \to 2$, $2 \to 4$, $3 \to 5$, $4 \to 3$ and $5 \to 1$. Inscribing this procedure in the template, we obtain the braid of this orbit, see Figure 2.10. The reader may want to verify that the orbit 00111 has the same braid, the only difference being that point 3 lies on the orientation reversing branch.

The relation between the kneading theory of unimodal maps and the associated braid structure has been explored in several works, beginning with the pioneer work of Holmes and Williams [Holmes and Williams 1985] where the horseshoe template was presented. Isotopic knots and their relation to bifurcation sequences were considered in [Holmes 1989], the same methods were taken into the "braids" presentation later, when conjugated braids associated with horseshoe braids were considered [de Carvalho and Hall 2003].

To decide whether two horseshoe orbits are conjugated or not is not a simple task. Since horseshoe orbits have only positive crossings in the standard template representation, and the number of crossings is invariant under conjugation, conjugated horseshoe braids must have the same number of strands (period) and the same number of crossings. For low period orbits, these rules shorten considerably the list of candidates. An exploration performed on horseshoe braids up to period eight [Mindlin et al.

---

[4]This concept comes from the study of 1-D unimodal maps. Recall that for "infinite contraction", the horseshoe map would behave as a unimodal map.

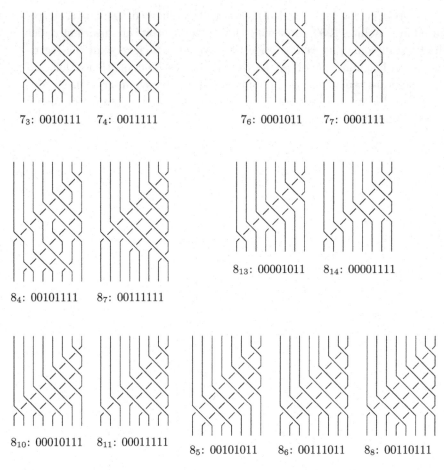

7_3: 0010111　　7_4: 0011111　　　　　7_6: 0001011　　7_7: 0001111

8_4: 00101111　　8_7: 00111111　　　　8_{13}: 00001011　　8_{14}: 00001111

8_{10}: 00010111　　8_{11}: 00011111

8_5: 00101011　　8_6: 00111011　　8_8: 00110111

Fig. 2.11　Sets of conjugated horseshoe braids (see text).

1993] indicates that the first set of candidates is of period seven, where there are two groups having two pairs of orbits each: $\{00111x1, 00101x1\}$ and $\{00010x1, 00011x1\}$ (the x stands for $0,1$ since each pair can be associated to saddle-node bifurcations in the high dissipative limit corresponding to one-dimensional maps). Considering orbits of minimal period eight, there are two saddle-node pairs with orbits presenting the same topological entropy, namely $\{000101x1, 000111x1\}$ and $\{000010x1, 000011x1\}$. There exists also a triplet of pairs formed by $\{001010x1, 001110x1, 001101x1\}$. These braids are displayed in Figure 2.11, where we exemplify taking $x = 1$ throughout.

A key element for the identification of conjugated braids was introduced in ([Holmes 1989]) observing that periodic orbits are invariant under the horseshoe map as well as under its inverse. The inverse of the (vertical) horseshoe map is another (horizontal) horseshoe map. Hence, forward and backward templates can be built and inversion symmetry can be constructed. Identifying pairs of braids conjugated by the symmetry allows the identification of conjugated orbits.

Chapter 3

Braids as Indicators of Phase-space Dynamics

Although knot and link information already say a great deal about the organization of a set of periodic orbits, we ended the previous Chapter noting that, for systems admitting a Poincaré section, the concept of braid gives a richer description. Let us continue the exploration of this topic.

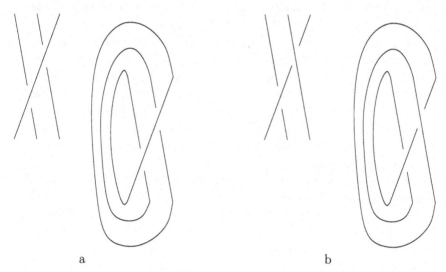

a   b

Fig. 3.1 Two period three orbits, shown as braids and knots. Orbit (a) corresponds to a rigid rotation while orbit (b) does not. You may want to compare the latter with the example in Figure 2.5.

Indeed, although both braids in Figure 3.1 correspond to period-three orbits, their properties are completely different. The one in Figure 3.1(a) (with braid word $\sigma_2\sigma_1$) can be thought as a periodic orbit of a rigid rotation of a disc[1], while the braid in Figure 3.1(b) (with braid word $\sigma_2^{-1}\sigma_1$) cannot. However, when regarded as knots, a few Dehn moves will convince the reader that both orbits are trivial knots. This observation makes it clear that whenever a global Poincaré section exists, the braid type of an orbit carries more information than its knot type [Solari and Gilmore 1988b]. We will heavily use this fact in the sequel.

3.1 Topological Entropy

The above example is less innocent than what it might appear. Indeed, the very existence of certain orbits in a continuous map of the disc forces the occurrence of other orbits. The "simplest" map of the disc that can host the orbit in Figure 3.1(a) would be the map that describes a rigid rotation by an angle of $2\pi/3$. All points in the disc except the origin belong to one such orbit, while the origin maps to itself being thus a period-1 orbit. Hence, all points in the disc are periodic and there are only two different classes of orbits for this rigid rotation map. On the other hand, we will realize later in this Chapter that the simplest map of the disc that can host the orbit in Figure 3.1(b) will display infinitely many different classes of orbits of infinitely many periods. Moreover, the growth rate of the logarithm of the number of classes with the period will be positive (roughly speaking, the number of (classes of) orbits grows exponentially with the period).

Finding one or the other orbit in a system allows to produce dramatically different predictions about the complexity (or rather the least possible complexity) of the problem.

A key concept to understand a map of the disc is then the number of topologically inequivalent periodic orbits for each possible period. This number can be heuristically associated to the notion of *complexity*. Simple maps have few orbits, complex maps have many. For rigid rotations, this number is bounded when regarded as a function of the period. Let us proceed with a definition that suffices for our purposes, although it may not be optimal. A deeper discussion can be found in [Fathi and Shub 1979, Katok 1980, Boyland 1984].

---

[1]We note on passing that a rigid rotation of the disc by an angle $2\pi/n$ (let $n > 1$ to avoid the trivial case), i.e., a map that to each point (r, θ) in the disc associates the rotated point $(r, \theta + 2\pi/n)$, generates a periodic orbit (or a class of homotopy-equivalent periodic orbits) of period n whose braid word has all the $n - 1$ generators just once and with positive exponent: $\sigma_1\sigma_2 \cdots \sigma_{n-1}$ (or some equivalent version).

Definition 3.1 **(Topological entropy)** Given an orientation preserving homeomorphism of the disc, we call *topological entropy* the quantity

$$h = \limsup_{n \to \infty} \frac{\ln N(n)}{n} \tag{3.1}$$

where $N(n)$ is the number of topologically inequivalent periodic points of period n.

For a rigid period-6 rotation, $N(n) = 7$ whenever n is a multiple of 6 while $N(n) = 1$ otherwise[2]. The definition above gives zero topological entropy. The horseshoe map, on the other hand, has 2^n periodic points of period n for each n and its topological entropy is $h_H = \ln 2$.

Zero topological entropy vs. positive topological entropy are qualitative measures of how much "nontrivial" dynamics a map has. This is in fact a central idea. For example, rigid rotations of the disc have zero topological entropy. More generally, for orientation preserving *homeomorphisms* of the disc if the map has zero topological entropy then every periodic orbit is *hereditarily rotation compatible*, i.e., it can be described as a composition of a number $m \geq 1$ of rotations built onto each other [Gambaudo et al. 1989] (the converse statement, namely that all maps displaying only hereditarily rotation compatible orbits have zero topological entropy, is also true for C^k *diffeomorphisms*, $k > 1$).

In the same spirit, Katok's theorem [Katok 1980] asserts roughly that a C^k $(k > 1)$ diffeomorphism (or some power of it) with positive topological entropy has an invariant set with at least as much structure as the horseshoe invariant set Λ. Hence, at least for sufficiently smooth diffeomorphisms, positive entropy corresponds to complicated ("chaotic") dynamics and zero entropy corresponds to simple dynamics.

3.2 Thurston's Theorem

Recall that our idealized problem situation is that the system under study admits a Poincaré first return map of which we only know a given finite set of periodic orbits. How can we get the most of this information?

---

[2]This may look like a weird way of counting orbits, since we count the period-1 *also* as a rather degenerate orbit of all multiple periods e.g., 31 or whatever. Using the *minimal period* the rigid rotation has of course just two orbits, but this apparent simplicity is overweighted by the fact that the definition of entropy has to be modified, since the logarithm does not admit zero as an argument.

First, consider that knowing the existence of some periodic orbits can be rephrased by saying that there exists a discrete finite point set P of the disc that is invariant under the action of the Poincaré map. In other words, the Poincaré map is an orientation preserving homeomorphism of $D\backslash P$ in itself[3]. We can further consider the homotopy class of orientation preserving homeomorphisms of $D\backslash P$ in itself. Our map is in that class and if we could say something general about the whole class, that would apply to our problem.

3.2.1   *Braid type*

Definition 3.2   We say that two homeomorphisms f and g of $D\backslash P$ in itself are *isotopic (rel P)* [Hall 1994a;b] if they are homotopic through homeomorphisms of $D\backslash P$ in itself. When the set P is understood, we just say "isotopic".

The isotopy equivalence class of a map f on $D\backslash P$ is called the *braid type*. We will see below that out of the action of the map one can produce a braid (or rather a braid word) describing the fate of the periodic set P under the action of f.

In particular, periodic orbits in $\mathbb{R}^2 \times \mathbb{S}^1$ correspond in this way to braid types. Moving the control section along θ one may cyclically reorder the letters of the braid word. In fact, given an arbitrary partition of a braid word W in two nonempty components: $W = W_1 W_2$, the braids represented by the words W and $W' = W_1^{-1} W W_1$ correspond to the same periodic orbit, for all possible choices of W_1. This partition can be operated by the homeomorphism induced by the flow when the dynamics evolves from one possible control section to another. Hence, the braid type takes into account the equivalence class of braids upon conjugation. This class is a topological invariant that gives a more detailed description than knot invariants.

3.2.2   *The theorem*

A crucial result, advanced by Thurston some 30 years ago that triggered a lot of research around this problem is the following one about the classification of surface homeomorphisms.

---

[3]It also maps P on itself and the whole D on itself, but since $D\backslash P$ has a richer structure than either D or P we may hope to gather more information regarding the map in this way.

Theorem 3.1 (Nielsen-Thurston) *Let Σ be compact and P a finite f-invariant set of points. Then f is isotopic to a homeomorphism g on $\Sigma \backslash P$ such that one of the following three cases occur:*

(1) g^n is the identity for some positive integer n (g is said to have finite order*).*

(2) g is reducible, i.e., there exists a g-invariant finite set of disjoint closed curves which are not boundary homotopic nor puncture homotopic in $\Sigma - P$.

(3) g is pseudo-Anosov.

The simplest homeomorphisms of a disc are rigid rotations as described above. In the reducible case, we can decompose P in a collection of two or more (irreducible) g^k-invariant sets. In fact, in the case that the points of P belong to just one periodic orbit, for some k, g^k maps each invariant curve onto itself and there are $l = p/k$ points of P within each curve. Hence, reducibility requires p not to be a prime number [Boyland 1984]. Confining ourselves to prime periods (or after decomposing reducible cases into the irreducible components) Thurston's theorem reduces to two alternatives: finite order or pseudo-Anosov homeomorphisms.

From the point of view of dynamics the last case in Thurston's theorem is the most interesting. In fact, pseudo-Anosov maps have many interesting properties that allow to assess a number of properties of the original (dynamical) map f. For the present purposes the three properties which are relevant are [Hall 1994a;b]:

(1) Let ϕ be a pseudo-Anosov homeomorphism on $D \backslash P$ that maps periodically the punctures of D (the unit disc) and let Q be a periodic orbit of ϕ with braid type γ and period q not lying completely in the border of D. Then, the number of periodic orbits with braid type γ and period q of any homeomorphism f in the isotopy class of ϕ is greater than or equal to the corresponding number for ϕ [Hall 1994a;b]. The result is not true for orbits lying completely in the border of D, but these are just a finite set. This means that since f and ϕ both present the same invariant set P and hence lie in the same class, f has at least the same number of periodic orbits as ϕ for each period $n \geq 1$ with the possible exception of the border orbits (which are a finite number of rigid rotations).

(2) The topological entropy of ϕ, $h(\phi)$, is a lower bound to that of f.

(3) Pseudo-Anosov maps admit a Markov partition from which $h(\phi)$ can

be computed (it is the logarithm of the largest-modulus eigenvalue of the associated Markov matrix) [Casson and Bleiler 1988].

The bottom line is that in the sense specified above, pseudo-Anosov maps have the least number of periodic orbits for all periods in their isotopy class. Hence, the strategy to follow in applications is the following: Given a Poincaré map and knowing one of its periodic orbits, compute the Thurston representative of the Poincaré map in its isotopy class. Apart from the reducible case, it is either a rotation or a pseudo-Anosov map. In the latter case, the original Poincaré map will have at least the same structure (periodic points, stretching, folding, linking, etc.) of the pseudo-Anosov representative up to isotopies. Hence, not only information about how "chaotic" our Poincaré map is, but also explicit information about which periodic orbits are necessarily present in our map can be recovered by identifying the pseudo-Anosov representative of our map in its isotopy class.

3.2.3 *Orbits That Imply Positive Topological Entropy*

From an application-oriented perspective, one would like to determine the conditions assuring that a given periodic orbit or its associated braid is compatible with a zero-entropy Poincaré map, or, alternatively, with a positive-entropy map. Beyond the zero-entropy result of the previous Section, Boyland [Boyland 1984] produced a recipe to determine whether certain orbits will imply positive topological entropy or not by considering the associated braid.

Take an irreducible braid, for example, the braid of a prime-period orbit (period n). Now sum the exponents of all the crossings σ present in the braid word. If this sum is not divisible by $n - 1$, where n is the period, then the braid implies (or 'has') positive entropy. If this is not the case, there is still a chance. If B^n, the nth power of the braid, is *not* a rigid rotation, then the braid also implies positive entropy.

Summarizing, the knowledge that the Poincaré map hosts some finite set of periodic orbits can be exploited as follows:

(1) Compute the linking properties of the set of orbits (as in the previous Chapter). This is a strong indicator since whatever test model, proposed set of equations, etc., describing our system can be safely discarded if the linking properties are not reproduced.

(2) Compute the braid of the orbit(s). Boyland's test as described above may decide whether the map is "chaotic" or not, i.e., whether the existence of the orbit(s) forces the Poincaré map to have positive topological entropy and an infinite number of periodic orbits which are not hereditarily rotation compatible.

(3) Compute a representative of the Poincaré map in its isotopy class (in the reducible case, decompose and compute representatives for all irreducible components). This representative is either finite order or pseudo-Anosov.

   (a) In the finite order case, the orbits are compatible with "simple" dynamics (hereditarily rotation compatible). Our specific system could be more complicated than that, but such information cannot be obtained via the given orbits.

   (b) In the pseudo-Anosov case a lower bound for the topological entropy can be computed and moreover, all periodic orbits present in the pseudo-Anosov representative will be present in the original map.

Note that Boyland's result is sort of a "quick test". One may decide whether the existence of an orbit necessarily implies that the Thurston representative of our map is pseudo-Anosov. On the other hand, the construction of the Thurston representative (be it pseudo-Anosov or not) or some equivalent object is a much more detailed tool, giving information about stretching, folding and periodic orbits of *all* periods. We leave the procedure of understanding the construction of finite order/pseudo-Anosov representative for a later Chapter, since it requires a lot of additional structure.

3.3   Highly Dissipative Systems

Another situation where a description in $\mathbb{R}^2 \times \mathbb{S}^1$ seems proper, is that of systems in higher dimensions where most of the dynamics is strongly dissipative. For example, when the dynamics can be separated in the following (or equivalent) terms:

$$x_{n+1} = f(x_n, y_n, \epsilon)$$
$$y_{n+1} = \epsilon g(x_n, y_n, \epsilon), \tag{3.2}$$

where $x \in \mathbb{R}^2$, $y \in \mathbb{R}^m$ for some integer $m > 0$, f and g are sufficiently smooth functions with $g(0,0,0) = 0 = f(0,0,0)$. In this situation, for

$\epsilon > 0$ sufficiently small, the Center Manifold Theorem can be invoked and the dynamics of 3.2 may be approximated by the dynamics on a globally attractive Center Manifold $y = h(x)$ [Natiello and Solari 1994]. Then we may formulate the following result [Natiello and Solari 1994]:

Theorem 3.2   *Let p be a finite union of periodic orbits of the above map in the center manifold and P the periodic orbits of the full map 3.2 dressed with their strongly stable manifold (i.e., the orbits decaying towards the center manifold exponentially fast). Then, $\pi_1(X_q) = \pi_1(Z_q)$.*

Here X_q represents the q-tuples of points in $\mathbb{R}^2$ where $x_i \neq x_j$ for $i \neq j$, being hence $\pi_1(X_q)$ the braid group of q strands; $\pi_1(Z_q)$ is the fundamental group associated to Z_q (Z_q is obtained "multiplying" each element of the q-tuples in X_q by $\mathbb{R}^m$). The index q labels the number of strands in p.

Chapter 4

Braids and the Poincaré Section

The presentation in the two previous Chapters demands for a synthesis of the different but related approaches we have followed. On one hand, the braid description of the periodic orbits of a flow in $\mathbb{R}^2 \times \mathbb{S}^1$, consists of "cables" (strands) given by the time-evolution itself. This description admits a nice graphical visualization. On the other hand, Thurston's theorem speaks about periodic points of a 2−d homeomorphism. The connection between both facts is that the periodic points in Thurston's theorem are the punctures produced by the strands of the braid on a control section.

The first question that arises is the following: Since Thurston's theorem does not know about the ultimate origin of the 2−d homeomorphism, but anyway deals with braid types, is it possible to establish the braid(-type) associated to a periodic orbit directly from the Poincaré section and first return map, without taking a "detour" through the flow?

The second issue goes closely into the "chaos" topic. Assume that we have a periodic orbit and braid type B, belonging to the dynamics of our system. Assume further that all eventual reducibilities have been cleared out and that our braid is hence pseudo-Anosov. Then the "simplest" map in the isotopy class of this braid type, meaning the map that has lowest number of periodic orbits for all periods in the class is the pseudo-Anosov representative. This rises many questions:

(1) How does the pseudo-Anosov representative map look like?
(2) Is there a simpler alternative map that has the same useful properties as the pseudo-Anosov? Which is in such case the difference between both maps?
(3) Given another periodic orbit of braid type B', is this orbit present among the set of orbits of the pseudo-Anosov representative for braid type B?

We will focus on these questions for the rest of this Chapter.

4.1 Braids on the Poincaré Section

Given a finite set of periodic points P on a Poincaré section, it is always possible to find a coordinate transformation (a homeomorphism) such that in the new coordinates the Poincaré section is the unit disc and the periodic points are lined up along a (horizontal) diameter (in the interior of the disc). In this situation we can give a consecutive numbering 1 to n to the periodic points in P.

Definition 4.1   (Line diagram) The line diagram is the union of the set P and a set of $n-1$ open straight line arcs going from element i to element $i+1$ in P, with $i = 1, \cdots, n-1$.

The line diagram is a connected portion of a (horizontal) straight line. Line diagrams have *edges* and *vertices* in the natural way.

Definition 4.2   (Circle diagram) A circle diagram is the Jordan curve obtained by adding to the line diagram a counterclockwise arc going from vertex n to vertex 1. See the leftmost diagram of Figure 4.1.

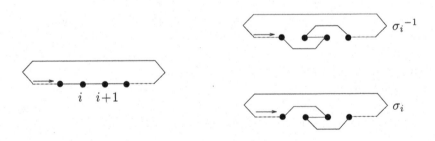

Fig. 4.1   The action of a 2−d homeomorphism on a circle diagram.

Circle diagrams are practical to relate periodic points to braids while line diagrams are useful to compute the topological entropy and an alternative to the pseudo-Anosov map. There is a one-to-one correspondence among both classes of diagrams, given by just adding or deleting the "closing arc".

With the help of circle diagrams, braids can be described graphically directly on the Poincaré section, without resorting to the flow in order to "read" the crossings of the threads [Natiello and Solari 1994].

There is however one piece of information from the periodic orbits that is lost when going to the Poincaré surface. In fact, the association between flows and Poincaré first-return maps is many-to-one. If the flow as a whole has a global torsion (i.e., it rotates as a whole around the flowing axis) which is an integer number times 2π, the first-return map remains unaltered. We will call these integer rotations a *full torsion* or a *full twist*. A flow compatible with a Poincaré map is called a *suspension*. A given Poincaré map admits many suspensions which differ from each other in the number of full twists.

The full twists constitute a subgroup Z_n of the braid group B_n. They are in fact the *center* of this group, i.e., those elements that commute with all elements of B_n [Natiello and Solari 1994]. Hence the quotient group B_n/Z_n is the relevant entity to characterize periodic orbits of any flow having a given first-return map. The connection between braids and circle diagrams is given by the fact that the equivalence classes of circle diagrams is in one-to-one correspondence with the quotient group B_n/Z_n between the braid group and the full torsions [Natiello and Solari 1994]. Related ideas in connection with line diagrams had been advanced without details in [McRobie and Thompson 1993].

In order to visualize this connection, take a periodic orbit and choose a circle diagram as a starting point. This implies having chosen an ordering of the periodic points along the circle. Now "slide" the circle along the flow until it returns to the control section as described in Figure 4.2. The resulting final circle diagram thus obtained will contain enough information to uniquely identify the braid of the orbit (up to global torsions) [Natiello and Solari 1994]. Each thread-crossing in the braid of the orbit corresponds to a topologically inequivalent deformation of the circle diagram. The elementary turns σ_i and $\sigma_i{}^{-1}$ are illustrated on the right diagrams of Figure 4.1. The braid word can thus be obtained by reading the elementary turns required to deform the original circle into the final one.

Hence, via the association of σ_i's to diagram deformations as suggested by Figure 4.1, the braids in B_n/Z_n can be directly read on the Poincaré section as deformations of a circle diagram. Acting on the initial circle diagram with the 2−d homeomorphism, the resulting Jordan curve differs from the starting curve in a number of elementary moves σ_i, produced in a given order. These elementary moves correspond to "letters" in the braid word of an element of the braid-type equivalence class B.

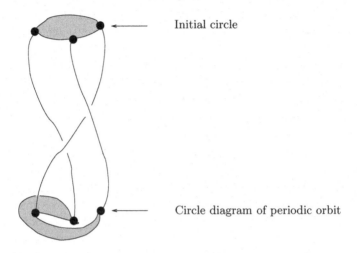

Initial circle

Circle diagram of periodic orbit

Fig. 4.2   Braids on the Poincaré surface: The image of the starting circle by the Poincaré map. Numbering the three invariant points from left to right along a counterclockwise circle, the associated braid reads $\sigma_2 \sigma_1^{-1}$.

4.1.1   *"Braidless" braids*

The dynamical description in terms of line diagrams, their images and the general properties of 2-D homeomorphisms of the disc can be considered in some sense (in some literary sense if the reader prefers it) *braidless*. The sense is that although finite invariant point sets of 2-D homeomorphisms can be associated to braids through clear and explicit rules, rather than the specific braid-group properties of the association, we are focusing on the mapping of edges (the lines joining consecutive vertices in the line diagram) by the homeomorphism as well as on other general properties of related homeomorphisms, together with the computation of other invariant point sets of that map related to the given one. In a strict sense, one cannot do without braids and still retain the richness of detail in the analysis that we will display in this Chapter. The procedure ends up identifying braid types and some properties of them that are relevant for a dynamical analysis. We will discuss other more strictly braidless approaches in terms of Homology groups in Chapter 6.

4.2   The Fat Representative of a Pseudo-Anosov Map

There is a rather natural way to produce a standardized 2−d homeomorphism from the line diagram and its image. A topological representative of the unit disc can be obtained by thickening the edges of the diagram to rectangles and the vertices to topological circles, in such a way that the resulting thickened diagram looks like a rectangle itself. A map on this rectangle lying in the same isotopy class of the original homeomorphism can be produced as follows:

- Topological circles map to topological circles in such a way that (i) P remains an invariant set of the map and (ii) the image of the set of circles lies in the interior of the set of circles.
- Thick edges shrink when necessary (by a sufficiently small factor $\alpha \in (0, 1/2)$) in the transversal direction (the thickened direction) and eventually stretch longitudinally having the image of the line diagram as a guideline.
- The resulting image object is a deformed rectangle, without self-intersections, that fits within the original rectangle.

We will call this map the *fat representative* of the original homeomorphism. It is illustrated in Figure 4.3, where we note that by adapting the homotopy we can make the map fit exactly within the original rectangle. This construction maps the interior of the unit disc one-to-one on itself, some connected arc(s) of the border are mapped two-to-one on the interior of the disc and the rest of the border maps one-to-one on the whole border.

The construction does not guarantee that the resulting map is pseudo-Anosov but helps a bit in that direction. Indeed, there is a problem, illustrated by the white portion of rectangle shown in Figure 4.3. This fat representative maps a small portion of the unit disc onto itself, stretched and folded as a horseshoe. In this case, the region coloured in white maps onto itself after four iterations. There are other small rectangular regions mapping in a similar way onto themselves.

If such a region involved in a horseshoe could be collapsed to a point, along with all its preimages, then we would produce a new map of the disc. For some of these regions, the collapse is possible without driving the new map outside the original isotopy class. The new map has a (significantly) smaller number of periodic orbits of all periods. This shows that the original fat representative was not even near to be a pseudo-Anosov map (recall that the pseudo-Anosov has the least number of periodic orbits for each period

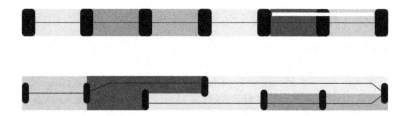

Fig. 4.3   A fat representative with a little horseshoe.

and the lowest topological entropy). But on the other hand it also gives the clue about how to improve the fat representative. If we systematically identify and collapse away all regions of phase space (in the complement of our periodic set) that will produce infinitely many periodic orbits while remaining in the given isotopy class, after this identification and collapsing process comes to an end we will have a fat representative in the same isotopy class of our original map but lying only a finite number of periodic orbits away from the pseudo-Anosov map of the class. Only that after all collapses the original line diagram may look very different.

This process can be quite involved and it has drawn the attention of a number of researchers during many years [Bestvina and Handel 1992, Los 1993, Franks and Misiurewicz 1993, Hall 1994a;b, Bestvina and Handel 1995, de Carvallo and Hall 2001, Solari and Natiello 2005]. Correspondingly, there exist a number of algorithms to produce the collapse or its equivalent. The bottom line is that allowing for more complicated diagrams than just line diagrams, each pseudo-Anosov isotopy class has at least one associated diagram whose fat representative has the same number of periodic orbits as the pseudo-Anosov map in the class, except for a finite number of orbits lying on the border of the topological disc. This gives a satisfactory answer to the first two questions posed at the beginning of the Chapter (it also contributes to the third question, which we address in the next Section).

4.2.1   *An algorithm*

In order to describe an algorithm based on the intuition of shrinking regions of the Poincaré section (i.e., of phase space) to a point, we start by generalizing the idea of line diagram to that of "tree".

A *tree T* is a finite connected set of vertices and edges without loops, such that each edge connects two vertices pairwise and edges do not intersect elsewhere than in the common vertices. In simplicial language, a tree is a connected finite 1−d CW-complex that does not contain a subset homeomorphic to a circle [Franks and Misiurewicz 1993]. Next step will be to adapt the idea of fat representative to trees.

Definition 4.3 (**fat representative**): Let $\widehat{T}$ be the topological disc obtained from a tree T by means of a suitable choice of "thickening" as described above (edges thicken to rectangles and vertices to circles). The *fat representative* $\widehat{\theta}$ of the homeomorphism F [Hall 1994b] is a map $\widehat{\theta} : \widehat{T} \to \widehat{T}$ with the following properties:

(1) $\widehat{\theta}$ is one-to-one and continuous
(2) $\widehat{\theta}(\widehat{T}) \subset int(\widehat{T})$
(3) $\widehat{\theta}$ coincides with F on P
(4) $\widehat{\theta}(T)$ is homotopically equivalent to $F(T)$ on $\widehat{T} - P$
(5) The image by $\widehat{\theta}$ of a fat vertex is contained in the interior of a fat vertex
(6) Given r belonging to an open edge of T, and calling π the projection sending $\widehat{T}$ to T, then for all t such that $\pi(t) = r$, $\pi(\widehat{\theta}(t)) = \pi(\widehat{\theta}(r))$ and moreover, $|\widehat{\theta}(r) - \widehat{\theta}(t)| = k|r - t|$, for some positive $k < 1$. k is constant on each open edge.

Recalling the example above, the need for removing (collapsing) regions of phase space arises only if the homeomorphism bends the image of the tree onto itself. We formalize this idea with the concept of fold.

Definition 4.4 (**Fold**): Let v' be a vertex of T and v the vertex of T which is the unique vertex preimage of v' by $\widehat{\theta}$. We say that θ has a *fold* f at v' whenever θ is not one-to-one restricted to any small neighbourhood of v. We say that $\widehat{\theta}$ has a fold at v' whenever θ has a fold at v'. We count one fold for every pair of contiguous edges at v with the same image by θ locally around v'.

We need to identify the preimages of a fold as well, since the relevant bending may not occur in the first iterate of F. Also, we need to decide which bendings actually require collapse, since it is the creation/elimination of periodic orbits, not the bending itself what is important. Following Hall [Hall 1994b] we will call the collapsible situations "bogus transitions". The following set of definitions will help us on the way.

Definition 4.5 (**Sector**): Let $\widehat{T}$ be the topological disc obtained from T by means of a suitable choice of π^{-1}. Consider the tree T as a point set imbedded in $\widehat{T}$. Every fat vertex of valence k (i.e., with k edges emerging from it) of $\widehat{T}$ is divided by T in k connected subsets that we will term *sectors*. The boundary of each sector contains only one vertex in T and portion(s) of edge(s) of T at that vertex. We will consider that the boundary belongs to the sector whenever necessary.

Definition 4.6   (Fold preimages): We define the set of *fold preimages* $PI(f)$ having sectors as elements as follows: $x \in PI(f)$, and in addition $y \in PI(f)$ iff $y \cap T$ maps (locally) one-to-one by θ^k onto $x \cap T$, for $k \geq 1$. Note that a sector cannot be associated to more than one fold and the sector at an endpoint cannot belong to $PI(f)$ since it cannot be mapped by θ one-to-one and onto the local part of T at a valence-m vertex with $m > 1$ in the way prescribed above. We will call $PI(\widehat{\theta}) = \cup_f PI(f)$, the set of all the sectors associated to folds in the map.

Definition 4.7   (Crossings): Consider an open edge e and its image by $\widehat{\theta}$. If we can divide e in three consecutive non-empty portions e_0, e_1, e_2 such that $\widehat{\theta}(e_i), i = 0 \ldots 2$ intersect three consecutive elements (sectors or edges) of the tree, we will say that $\widehat{\theta}(e)$ **crosses** the second intersected element (the one corresponding to e_1). Notice that if $\widehat{\theta}(e)$ crosses an edge, the edge portions e_0, e_2 intersect sectors, since edges connect sectors.

Definition 4.8   (Bogus Transition): Consider the set of *fold crossings* $CR(\widehat{\theta})$ indicating which sectors or unions of consecutive sectors associated to the points P are crossed by the image by $\widehat{\theta}$ of an edge of T. The orbit by $\widehat{\theta}$ of the elements in CR consists of a sequence of sectors or union of consecutive sectors which could either map into one or more folds in a finite number of steps or be infinite. In the same way, the orbit by θ of the border of these sectors in T either is 2-to-1 after a finite number of steps (in which case we say that the orbit *terminates* in the fold) or keeps being 1-to-1 for any number of iterates. We say that the tree T has a *bogus transition* at all the folds lying in the forward image by $\widehat{\theta}$ of an element of $CR(\widehat{\theta})$ whose orbit *terminates*, in the present sense.

For each fold f with a bogus transition, the set $BT(f)$ is defined as the subset of $PI(f)$ that has nonempty intersection with the forward image of the elements of $CR(\widehat{\theta})$. $BT(f)$ indicate the sectors where tree modifications will be necessary.

The existence of bogus transitions (it is in fact a special case of them called *recurrent* [Solari and Natiello 2005]) motivates the need for modifying phase space. Next we have to specify exactly how to perform the collapse. Intuitively, too small a collapse will still require further collapse, while too large a collapse may create spurious orbits. The goal is to eliminate regions of phase space that carry a significant number of periodic orbits without adding new orbits, in order to get a new phase space (a new tree) and corresponding new map $\widehat{\theta}$ in the isotopy class of F but with less periodic orbits. We will specify the relevant "size" of the collapsing areas with the following definitions, ending with a precise definition of the concept of collapse.

Definition 4.9   (Preimage of a fold): Let $\widehat{\theta}$ have a fold f at v'. The two (adjacent) folding edges at the point v, unique vertex preimage of v', define two branches on the tree T.

Consider the sector $x(f)$ associated by $\widehat{\theta}$ to the local interior of the fold discussed in the definition of fold. Let $A(f)$ and $B(f)$ be the extreme points of the arc belonging to the border of the fat tree at the sector $x(f)$, $\partial\widehat{T} \cap x(f)$. Further consider $\alpha(f) = \pi(A(f))$ and $\beta(f) = \pi(B(f))$ and the transversal arcs $A(f) - \alpha(f)$ and $B(f) - \beta(f)$. We have that $\theta(\alpha(f)) = \theta(\beta(f))$.

The connected region limited by the arc in $\partial\widehat{T}$ connecting $A(f)$ and $B(f)$ through the fat-vertex v, the transversal arcs $A(f) - \alpha(f)$, $B(f) - \beta(f)$ and the tree, T, will be called a *preimage of the fold, $PF(f)$*.

The region $PF(f)$ can be extended by monotonously moving the points $A(f)$ and $B(f)$ on $\partial\widehat{T}$ in opposite directions as long as the following requirements are satisfied:

(1) $\theta(\alpha(f)) = \theta(\beta(f))$
(2) $\widehat{\theta}(\partial\widehat{T} \cap PF(f))$ can be deformed into a portion of a segment transversal to the tree at $\theta(\alpha(f))$

Any such region will be as well called a preimage of the fold. In particular we will be interested in the largest possible region of this kind, which we call $MPF(f)$, the **maximal preimage of the fold.**

Definition 4.10   (Crossing a PF): We will say that the image of an edge e crosses $PF(f)$ whenever there are two points in e, e_A and e_B defining a portion of an edge $e_2 = [e_A, e_B]$ and such that $\theta(e_A) = \alpha(f)$, $\theta(e_B) = \beta(f)$ and $\widehat{\theta}(e_2)$ is homotopic in $\widehat{T} - \{V\}$ to $PF \cap \partial\widehat{T}$, keeping $\theta(e_A)$ and $\theta(e_B)$ fixed in the homotopy.

Continuing with the discussion of requirement (2) above for extending the preimage of a fold, it is worth to render its motivation clearer. Suppose that for some integer n and edge e, $\widehat{\theta}^n(e)$ crosses PF, then $\widehat{\theta}^{n+1}(e)$ will map across the fold region in the same way as $\widehat{\theta}(\partial\widehat{T} \cap PF(f))$. If this image is homotopic to a transverse arc, it will disappear via a suitable homotopy when $\widehat{\theta}^{n+1}$ is pulled tight, however, if there are "obstacles" in the form of vertices (added or original vertices) such homotopy cannot exist.

When the map presents a single fold (as in the motivating discussion above), the MPF is easily identified. However, when more than one fold is present in a map, the folds may have adjacent prefold regions [1]. Under such circumstances it is possible to make further identifications considering simultaneously all the folds of the map.

Definition 4.11   (Collapse of a fold): Let $\widehat{\theta}$ have a fold at v'. The collapse of a fold consists in identifying points in $\widehat{T}$ in such a way that α and β coincide and PF has empty local interior. We call the identified *end* point $*v = \alpha = \beta$.

Definition 4.12   (Collapse of a bogus transition): Let $\widehat{\theta}$ have a fold at v'. The *collapse of a bogus transition* consists of (a) the simultaneous collapse of disjoint regions (with the exception of at most a common endpoint for adjacent regions) around all the preimage sectors of the sector at v involved in a (recurrent) bogus transition (i.e., the set $BT(f)$) and (b) the collapse of interior portions of the edges that map by θ^k on the collapsed regions.

The collapse generates extra vertices not present in the original invariant set. The fate of these vertices will depend on each specific situation. They may build another invariant set disjoint with the original one, or disappear at some step of the procedure (they coincide with preexistent vertices). At intermediate steps, the added vertices may be eventually periodic, i.e., there is a proper subset of them that is periodic while the remaining other vertices eventually map by $\widehat{\theta}$ on this subset. Such a situation indicates [Solari and Natiello 2005] that the collapse procedure is not yet finished.

Definition 4.13   (Exhaustion of a fold): A fold is *exhausted* whenever the fold has empty local interior. We also say that the fold is *partially exhausted* if there is at least one collapsed region, say $j < n$, such that it is the maximal j-preimage of the fold, $PF_j(f)$. A partial exhaustion implies the presence of other folds.

The following lemma specifies how the collapse is to be performed.

---

[1] By adjacent we mean that $A(f) = B(f')$ or $A(f') = B(f)$, i.e., we are not considering as adjacent two regions which lie at different sides of a common edge.

Lemma 4.1   *The collapse of a bogus transition can be increased without creating new folds until one of the following situations arises:*

(1) Two adjacent collapsing regions have one endpoint in common

(2) All added vertices are in coincidence with preexisting vertices and the bogus transition no longer exists

(3) All added vertices are in coincidence with preexisting vertices and the number of explicitly collapsed regions needs to be increased to continue the collapse

(4) The fold is partially exhausted

(5) The fold is exhausted

The above situations are the basis to define a *collapsing step*, i.e., to collapse (when necessary) as much as it is allowed by the previous lemma. Within this framework, the algorithm to produce a tree and associated fat representative having the same number of periodic orbits for each period as the pseudo-Anosov representative except for a finite set of border orbits reads as follows:

Algorithm

(1) Identify all folds in the map.
(2) Detect folds with recurrent bogus transitions. If there are no recurrent bogus transitions end.
(3) Select the fold with eventually periodic added vertices associated (if there is any) or a fold with (recurrent) bogus transitions otherwise. If no fold can be selected end; otherwise collapse the bogus transition or fold:

    (a) Mark regions to be collapsed adding valence-3 stars at points of BT.

    (b) Perform one collapsing step.

    (c) Eliminate cycles among edges with at least one star as endpoint collapsing the edges to a point.

    (d) Go to (3).

Further details on the algorithm, its motivation, uses, and some examples can be found in [Solari and Natiello 2005]. For the case shown in Figure 4.3 after a few collapsing steps sketched in Figure 4.4 we obtain the final tree and its fat representative at the bottom of the figure.

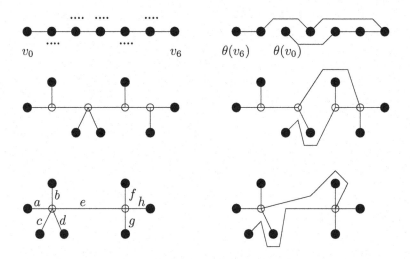

Fig. 4.4   The period-7 orbit of Figure 4.3 revisited. Dots in the first row indicate regions of $PI(f)$.

4.3   Trees, Topological Entropy and Orbit Forcing

Trees and fat representatives without bogus transitions are a powerful tool to compute the topological entropy. In fact, edges serve as a Markov partition, or as symbols for a symbolic labeling of the orbits (in line with the 0 and 1 labeling associated to the horseshoe shift map) and periodic orbits of period k may arise when an edge maps on itself after k iterations. A Markov matrix R for this partition can be computed, assigning to element R_{ij} the number of times that edge j maps on edge i after one iteration of the fat representative. The logarithm of the largest eigenvalue of this square matrix yields the topological entropy. For the previous example, with the edge labeling shown in Figure 4.4, the matrix reads

$$
\begin{pmatrix}
0 & 0 & 1 & 2 & 1 & 0 & 0 & 0 \\
0 & 0 & 0 & 0 & 0 & 1 & 0 & 0 \\
0 & 0 & 0 & 0 & 0 & 0 & 1 & 0 \\
0 & 0 & 0 & 0 & 0 & 0 & 0 & 1 \\
0 & 0 & 0 & 0 & 1 & 2 & 0 & 0 \\
0 & 0 & 0 & 1 & 0 & 0 & 0 & 0 \\
0 & 1 & 0 & 0 & 0 & 0 & 0 & 0 \\
1 & 0 & 0 & 0 & 0 & 0 & 0 & 0
\end{pmatrix}
\tag{4.1}
$$

and the topological entropy is $h_T = \ln(1.61094) = 0.47682$ (rounded up to the fifth figure).

4.3.1 *Orbit forcing*

We are now in a position to discuss the final question posed at the beginning of this Chapter. Given a second periodic orbit (i.e., given the period and the braid type of an orbit other than the initial one), is this orbit going to occur in the *minimal* fat representative (i.e., that obtained with the algorithm above)? Also, which orbits do occur along with the original one?

Considering our initial example, apart from the period-2 orbit given by the two added vertices, for a periodic orbit to occur in the fat representative, it is necessary that the edges where the points of the orbit belong are mapped periodically onto each other by the fat representative. The sequence $a \to d \to h$ allows for the existence of period-3 orbits. The other possible sequences of mapped edges are $a \to c \to g \to b \to f \to d \to h$ allowing for period-7 and $a \to [e \cdots e] \to f \to d \to h$, where the bracket indicates that since edge e maps onto itself, this family of sequences allows for orbits of period-5 and any higher period as well. The braid types of these orbits can be computed from the tree and its image (after more or less involved manipulations).

Hence, the computation of the minimal fat representative gives an answer to the question of *orbit implication* or forcing, stated in the following terms: Which orbits are necessarily present on a 2−d homeomorphism along with a given one? First, we note that the orbits that necessarily will be present in any map in the isotopy class of the given one are those occurring in the pseudo-Anosov representative. The answer that this method provides is that the minimal fat representative has the same orbits as the pseudo-Anosov representative except for a finite set of orbits lying in the border of the topological disc. Hence, the forced orbits can be read from the final tree.

4.3.2 *Orbit pruning*

In the light of our exposition, having a finite set of periodic orbits obtained via analysis of experimental data, we are far from having a fully-specified dynamical system. The procedure described above, generates the (in some sense) "simplest" dynamical system that can host our orbits. This simplest map can be regarded as a member of a larger family, which loosely speaking can be defined as the family of all maps that can host our data orbits. However, because of other experiences imbedded in our research programme, sometimes it may be convenient to pick up a special map present in the family. It might be e.g., that for reasons not exposed in this book we conjecture that the originating map is hyperbolic, and then we select *the* hyperbolic member (or *one* hyperbolic member) of the family.

For example, the periodic orbit in Figure 4.4 can be regarded as a horseshoe orbit since its braid type appears in the horseshoe map. A way to identify the orbits forced by the presence of the period-7 is to indicate the complement in the set of horseshoe orbits of the forced-orbits set. This is, the orbits that we would have to trim or prune away if we begin with the horseshoe map and end with only the orbits implied by the period-7. It is in this sense that the problem labeled *orbit pruning* [de Carvallo and Hall 2001; 2002] focuses in the relation between the remaining periodic orbits after the collapse and the periodic orbits of the "original object" (usually a map with some specific property, in this Section it will be the horseshoe map).

Carvalho and Hall formulate this question as: Given a reference homeomorphism F (the "original object") in the family of maps hosting our periodic orbits, we regard the dynamics of the maps in the family as the dynamics of F less *that which is pruned away* [de Carvallo and Hall 2001]. The collapse "prunes away" orbits and the "last" element in the family is the minimal fat representative.

Intuitively (and actually) the gluing ([Franks and Misiurewicz 1993]) or collapse ([Solari and Natiello 2005]) ideas put in act some sort of "pruning" in the sense that different regions of phase space are identified and/or collapsed to a point and along with them large sets of periodic orbits are identified or collapsed. References [de Carvallo and Hall 2001; 2002] extensively discuss the collapsing process, giving a content to the concept of "pruning" in terms of modifications in phase space and their consequences for the set of orbits.

The uneasiness that transpires from these paragraphs comes from the original assumption, i.e., that we assume that a given homeomorphism F is our *reference map*. We know only of a finite set of braid types, nothing more. To assume a specific reference map is not supported (nor is it necessary) by the considerations presented so far in this book. Whether it is a sound assumption or an irrelevant one, needs additional external support. It might be better or worse motivated in different specific cases.

To render the ideas behind pruning explicit, let us discuss the period-7 as a horseshoe orbit. We begin by constructing a sort of fat representative for the horseshoe, distinguishing an eventually periodic orbit of (eventual) period-1 and three subsets of the disc $\pi(V_0), \pi(V_1), P1$ where $\theta(V_0) \subset V_1$ and $\theta(V_1) \subset P_1$. The original sets V_0, V_1 and P_1 are given in Figure 4.5. The orbit $\pi(P_1)$ is not in the horseshoe and is the characteristic extra-boundary orbit of the fat representatives.

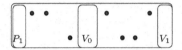 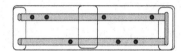

Fig. 4.5 Fat representative of a horseshoe map and period-7 orbit. Left, the fat tree (a line diagram in this case) with fat eventually periodic orbit $V_0, V_1, P1$ and the dots corresponding to the period-7 orbit. Right, the image by θ of the tree.

Identifying the left side of the horseshoe with the label 0 and the right side with the label 1, the leftmost point of the period-7 reads in terms of symbolic sequences of the horseshoe as $(00101X1)^\infty$, where $X \in \{0, 1\}$ (in the figure $X = 1$). The 0's and 1's in the symbolic sequence indicate which side of the horseshoe is visited successively. Our first task will be to trim the orbits to the right of the rightmost point of the period-7, i.e., orbits containing sequences higher than $(100101X)^\infty$ will be pruned away as well as their preimages (higher is taken in the sense of symbolic sequences of unimodal maps [Collet and Eckman 1986, Solari et al. 1996a]). This first trimming takes us to the first line of Figure 4.4. Notice that points to the left of the fourth periodic point (v_3 in Figure 4.4, the preimage of the fold) will carry horseshoe names beginning with 0 while points to the right of the v_3 will begin with 1.

When further elimination of little horseshoes proceeds, going from line one to line two in the diagrams of Figure 4.4, the edges carry their own label inherited from the horseshoe, except for the case of the edge d which is the result of collapsing regions with different label. Hence, the horseshoe-genealogy of this edge is ambiguous and we will label it X. In the last step of Figure 4.4 no essential alteration of the labels is produced and hence, the following assignment of inherited edge-labels occurs: $a, b, c \to 0$; $d \to X$; $e, f, g, h \to 1$. In this way, the periodic orbits forced by the period-7 horseshoe orbits of Figures 4.4 and 4.5 correspond to the sequences: $e^\infty \to 1^\infty$, or periodic repetitions of the syllables $acgbfdh \to 00101X1$, $ae^k fdh \to 01^k1X1$ and $adh \to 0X1$ (with $k \geq 1$).

Since the new tree (and edge-structure) has been derived from the starting one by means of reproducible operations[2], the process arriving to the final tree can be recasted as (a) produce a finer Markov partition than the horseshoe one (compatible with the given orbit and reference map) and (b) identify certain units of this finer partition with each other and collapse other units to a point. In this way, the periodic orbits hosted by the minimal fat representative can be recasted in the original (horseshoe) language and are readily identified as a pruned subset of the horseshoe orbits.

4.4   Examples

4.4.1   *First example*

In Figure 4.4 we display a chaotic period-7 orbit. Each row of the figure displays T along with $\widehat{\theta}(T)$. The different rows are produced after successive applications of steps of the algorithm. Added vertices have white colour.

Labeling the vertices v_i, $i = 0, \ldots, 6$ from left to right, we see from row 1 that there is one fold (hence necessarily at an end point, $v_6 = \widehat{\theta}(v_3)$). The dotted regions above (U) and below (D) points of T denote the regions that map on the fold by the iterates of $\widehat{\theta}$, the sequence defining PI is then

$$2U \to 5U \to 1D \to 4D \to 3U \to *f = v_6.$$

We have that $CR = \{2U, 3D, 4U, 5U\}$ and since $2U$ is in CR, we have that $PI = BT$. The set PI contains all elements of the above sequence up to (and except) the fold v_6.

Pieces of $\widehat{\theta}(T)$ passing above or below the dotted vertices indicate the existence of bogus transitions. Collapsing around the dots produces five added vertices $*2 \to *5 \to *1 \to *4 \to *3$ which eventually become four after collision of the preimage of $*f$ (namely $*3$), and its contiguous star, $*2$, (row 2). Further, the two outermost edges having added vertices as endpoints "verified" map onto each other and can be collapsed. The resulting diagram still has a fold but no bogus transition.

4.4.2   *Second example*

Let us now turn to the example in Figure 4.6. We label the vertices from left to right as 0, 1, 2, 3, 4, letting the unaligned vertex be number 3. We label

---

[2] For instance, edge b in Figure 4.4 consists of contiguous portions of the edges adjacent to the second vertex from the left in the original line diagram (i.e., v_1), identified by the collapse.

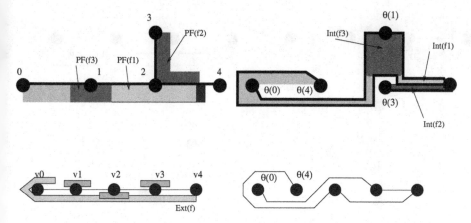

Fig. 4.6 Prefolds and extended preimage of a fold. The example presents three folds, two of them are adjacent. In the second line, the tree after eliminating the bogus transitions.

the sectors at each vertex as U, D, L or R (up, down, left, right) as suits the natural orientation of the Figure (vertex 2 has sectors L, R and D but no U-sector). Finally, label the edges from left to right as a, b, c, d, being c the vertical edge. $CR = \{0, 1D, 2L, 2R, 2R + 2L\}$ while $PI(f_1) = 2D$, $PI(f_2) = 2R$, $PI(f_3) = 1D$, for the three folds indicated in the figure. Among the elements of CR only $1D$ and $2R$ have finite orbits, all others, or their images, are the only sector at an endpoint. Hence, $BT(f_2) = PI(f_2)$, $BT(f_3) = PI(f_3)$ since the corresponding PI's are subsets of the set of elements of CR having finite orbits. On the other hand, $BT(f_1) = \emptyset$ and f_1 has no bogus transition.

Regarding f_2, we have that $v = 4$. For $q = 1$, $v_1 = 2$ and $\widehat{\theta}(b)$ crosses $2R$, where b is the edge between vertices 1 and 2. As for f_3, we have that $v = 3$. For $q = 1$, $v_1 = 1$. $\widehat{\theta}(a)$ and $\widehat{\theta}(d)$ cross $2D$.

The resulting tree without bogus transitions is shown in Figure 4.6.

4.4.3 *Third example*

Next, we consider the case of Figure 4.7.

Vertices and interesting sectors are labeled in the first row of the figure. Name the edges as a, b, c, from left to right. There is a fold at $v = 4$ with $CR = \{2D, 3D, 3U, 1\}$ and $PI = \{3U, 2D\} = BT$ (trivial).

After collapsing we arrive at the figure shown in the second row of Figure 4.7, with four edges and five vertices. This first collapsing step

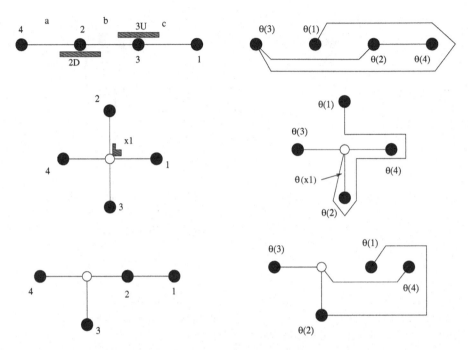

Fig. 4.7   Collapsing regions at opposite places at a vertex First line left: the extended preimage of the fold, and preimages PF_2 and PF_1 of the fold to be collapsed. Right: the image of the tree (solid line) and the image of $\widehat{\theta^2}(c)$ (dotted line). Second line: After the first collapse the bogus transition persists but the collapse proceeds only at the first preimage of the fold. Third line: the fold persists but there is no bogus transition any more.

ended with a partial exhaustion of the fold when the two added collapsing regions at opposite sides of an edge are in contact. Labeling the sectors at the period-1 added vertex x_1, x_2, x_3, x_4 in counterclockwise order starting from the preimage of the fold ($x1$) we can see that $CR = \{x_1, x_4, 3, 1\}$ and $PI = \{x_1\}$, hence, there is a bogus transition.

The final step is taken collapsing at $x1$ until the bogus transition is eliminated when the region of collapse reaches 2. The remaining tree has a fold but no bogus transition.

4.4.4   Fourth (last) example

We conclude the examples Section by considering a case shown in [Franks and Misiurewicz 1993], which we display in Figure 4.8.

There are two folds, one at $5 = \widehat{\theta}(4)$ which we call fold $f1$ and fold $f2$

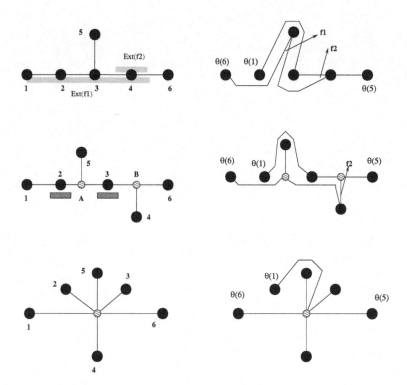

Fig. 4.8   A revisited example. In the first step the fold $f1$ is eliminated in a perfect exhaustions identifying a period-two orbit $\{A, B\}$. In the second step the collapse is performed under 2 and 3 eliminating the bogus transition at $f2$.

at $4 = \widehat{\theta}(3)$. $PI(f1) = \{3L, 4U\}$ while $PI(f2) = \{3D\}$. $CR = \{3L, 3D + 3L, 5, 2D\}$ (U, D, R, L indicate above, below, right and left respectively, as mentioned above), $BT(f1) = \{3L, 4U\}$ and $BT(f2) = \{3D\}$ (note that $3D$ is in the orbit of $2D$). The bogus transitions are eliminated after two steps, yielding the third row of the figure. In the first step the fold $f1$ is exhausted leaving a period-two orbit behind (perfect exhaustion). In the second step the bogus transition at $f2$ is eliminated, the fold moves all the way to A passing first through B and an added vertex remains under A (produced by drawing *2 and B together in the collapse). The edge connecting both stars maps onto itself and can be eliminated by item (3c) of the algorithm, leaving behind a period-1 added vertex. Note that in this second collapse the set $PI(f2) = \{2D, 3D\}$ contains two elements, one of them $(2D)$ was not present in the previous analysis. The sets $BT(f)$ and $PI(f)$ may change after a step is performed.

Chapter 5

Reconstruction of Phase-space Dynamics – Basic Course

5.1 Introduction: Naive Measurements

Starting with this Chapter, we are going to address a radically different class of problems. While up to now we have dealt with mathematical results aimed to deepen the understanding of the consequences of having an (unknown) 2−d homeomorphism hosting a (known) finite set of periodic orbits, sooner or later we will be forced to leave those safe waters to enter the insecure ground of modeling and interpretation of experimental data.

Within the limits of this manuscript, the interpretation of experimental data will in the end amount to modeling and identifying periodic orbits in 3-space and following the consequences of this identification for the understanding of the underlying problem. We will approach these questions in smaller steps (as usual), i.e., first via a naive measurement theory and later via actual (experimental) measurements. We repeat the warning of Chapter 1: There exist no black-box methods producing unambiguous passive understanding of experimental data without additional efforts from the researcher. You have to critically consider what you are doing.

Given a dynamical system $\dot{x} = f(x)$, where f and x are of the sort we have studied in the previous Chapters, we define a *measurement* as a smooth function $y = \phi(x, w)$ where y is a real number and w are extra variables describing the state of the measuring device. A *naive measurement* is that where $w = g(x)$ (g also smooth enough), i.e., where the state of the measuring device is completely specified by the dynamical system. The measurement functions are in practice far from arbitrary, but they may look very differently depending on each specific setup.

More realistic measurement theories may be obtained by refining the definitions above. For example, the outcome of the measurement could be taken to belong to a closed and bounded real interval $[x_0, x_1]$, or even to a discretization of this interval, i.e., the set $x_0 + th$ where $h = (x_1 - x_0)/M$, $t = 0, \cdots, M$ and M is a "large" positive number (say $M = 2^N$, where N is the resolution in bits of the measuring device, nowadays $N \approx 32$ at best, see the next Section). Moreover, the measuring device may have its "own life", with a behaviour depending on other things than just x (here represented by the w's).

5.1.1   *Time-series*

What the scientist at best has access to is a time-series of y-values. A time-series is just a string of numbers emerging consecutively out of the measuring device. The choice of measuring times is an additional problem. Whenever there is no natural way to decide when to perform the measurements, we will assume that the time-interval between consecutive entries in the time-series is constant (and much smaller than the typical dynamical times involved).

Naive measurement theory is a difficult enough starting point. Even with an almost trivial function $\phi(x, w) = x_1$ (i.e., the first component of x in some suitable coordinate choice) the task of reproducing x and f out of a time-series of x_1 is enormous.

Two observations are proper at this point. Given a time-series it may happen that its outcome does not fit within any reasonable tolerance bounds in the narrow costume developed in the previous Chapters. In such a case there is nothing left but accepting the fact and seeking other ways/models/methods to interpret the data. Second, and equally important, all interpretation of experimental data is provisional, i.e., it holds as long as (a) it is internally consistent, (b) it is the best at hand (i.e., the one that explains previous observations and predicts future observations the most accurately) and (c) it is refutable (it can be put to test and criticism).

5.2   Data Analysis

5.2.1   *Filtering and interpolation*

Despite the fact that we believe that an accurate dynamical description of our problem can be achieved with differential equations describing the evolution in continuous time, measurements usually come out in discrete form. Not only a discrete set of measurements, but also discrete outcomes. The roughest discretization is the "present-absent" pair (one bit). In general, digital measuring devices produce integer outcomes (after rescaling and shift of origin) in the interval $[0, 2^N - 1]$, where $N > 0$ is the accuracy in bits of the measuring system. For orientation, MIT's arrhythmia database [Narayanan et al. 1998] (see also www.physionet.org) which is about 35 years old, has $N = 10$, standard stereo audio quality can be achieved with $N = 16$, but even $N = 32$ is already built-in in cheap PC audio cards.

If the signal was a real-valued function and the measurement is integer-valued, then a time-series consists of a combination of dynamical information and some uncertainty (hopefully small, but in any case at least the uncertainty because of roundoff will be present). This uncertainty is usually called "error" (although this does not necessarily mean "mistake"). To filter away the error can be a sound practice. In fact, we can hardly recall research reports where raw data subject to measurement error is used without any error-filtering at all. Unfortunately, filtering has to be done blindly. There are no general rules to be followed; it is part of the research work to understand what to filter away and how to do it.

Even assuming error-free data, there is another unavoidable problem in dealing with time-series. The data will have *false* maxima and minima. In other words, the local maxima and minima of our real-valued continuous variable $x(t)$ will not coincide (neither in position nor in amplitude) with those of the measured outcome (which at best will yield values of the form $y_k = \text{rint}(x(t_k))$, where rint represents the integer roundoff or truncation of the measuring device). Whenever the problem allows for specific corrections, it is a sound policy to implement them.

For example, if the data consists of narrow, high, sharp peaks followed by silent periods of essentially zero output (as can be observed in lasers [Solari et al. 1996b]) it seems reasonable to filter away the silent periods to zero and to interpolate the position and height of the maxima using the measured data around each maxima via e.g., Lagrange interpolation.

5.2.2 Close returns

The first thing to do in order to apply the periodic-orbit theory developed previously is to identify periodic orbits within a time-series. A time-series is likely to present some portions where a nice and almost periodic recurrence is observed (with some errors coming from different sources), separated by other portions having no apparent periodicity.

The interpretation of experimental data rests on the assumption that these nearly periodic data portions are actually blurred versions of unstable periodic orbits of the system (they *must* be unstable, otherwise once the orbit is sufficiently close to a stable periodic orbit, it will never depart from it; the whole time-series would ultimately consist of just one periodic orbit following a more or less short transient). Since there is some error involved, the detection of periodic orbits is always "provisional" and valid within some error bounds.

The method of *close returns* [Lathrop and Kostelich 1989] considers that a portion of a trajectory belongs to a periodic orbit whenever for some $p > 0$, $\epsilon > 0$ sufficiently small and a fixed nonnegative integer P,

$$POC_{ip} = \sum_{j=0}^{P} |x_{i+j} - x_{i+p+j}| < \epsilon. \tag{5.1}$$

The data points $\{x_i, \cdots, x_{i+p-1}\}$ are regarded as a "periodic orbit candidate", for an orbit of (not necessarily least) period p. This is the period in the time-series, i.e., p time-steps. Its period in clock-time will be $T = ph$ where $h = (t_1 - t_0)/(N - 1)$ is the time-interval between recordings and N is here the total number of data points. The identification of periodic orbits will rest on some form of convention concerning which P is large enough and which ϵ is sufficiently small. Because of the various sources for differences within the data, P cannot be very large, and it will tend to deteriorate for large values of p.

5.2.2.1   *Guided example*

Let us illustrate the technique with an example.

(1) Write a programming code that numerically solves the Lorenz equations.
(2) Generate a time-series by recording the y-coordinate of the above solution evenly in the time interval $t \in [t_0, t_1]$. Take a total of $N = 2^{14}$ points (or more if you want), which we call $y(i)$.
(3) Digitalize the output, with 14 bits (or something else if you want) i.e., compute the maximum r and minimum s of the data set and rewrite: $v(i) = rint((2^{14} - 1)(y(i) - s)/(r - s))$, where we have rounded up to the nearest integer. Now we have an integer data set in the (closed) interval $[0, 16383]$.
(4) Check for close returns. Write a program that for a pair (i, p) computes the difference POC in Eq. (5.1). In the Lorenz system, with N chosen

as above, $t_0 = 115$ and $t_1 = 185$ in the standard time-units of the Lorenz equations, a low-period orbit will have a few hundreds of data points, so let us perform the check for about $P = 384$ points (so that a considerable portion of a whole revolution is still a close return). Note that a low value of P may give false close returns, while a large value of P may give no close returns.

(5) Choose your periodic orbit candidates among those (i, p) that give you "low" POC values. A possible criterion is to take a POC that yields an average relative error per point well below 0.01, i.e., $POC < ERR <<$ $P \cdot (2^{14}) \cdot 0.01$. To fix ideas we took $ERR = P \cdot (2^5)$. The candidate orbit starts at i and keeps going for p data points. In the general case, how to choose ERR is a matter of experience.

(6) For comparison, generate another array x where you pick the elements of v randomly and repeat the close returns search for x.

The outcome of this example was two periodic orbit candidates, with a relatively large degree of repetition (the pair (i, p) and also the pairs $(i + 1, p)$, $(i + 2, p)$, ... were detected as candidates, for many consecutive starting points) which means that the candidate orbit was "good" for more points than just P. Increasing P decreases the repetition. As a comparison, the set x having no Lorenz-dynamics whatsoever goes through the same tests yielding zero periodic orbit candidates. It would be highly surprising otherwise, since we have actively destroyed the dynamical information when building x.

The question that the close returns may have arisen is what sort of periodic orbits one has obtained. Perhaps the intuitive thing to do was *first* to generate some model dynamical system out of our scalar time-series and *afterwards* search for periodic orbit candidates in the generated system. Equation (5.1) to find periodic orbit candidates can in such a case be used almost as it is, only that an appropriate distance between the generated multidimensional data points has to be used instead.

We may claim that if a consecutive string of p points in the original data set is *not* a good close return candidate, then it will not be it either after having transformed the data into a d-dimensional system. Perhaps the generated dynamics suggests us to discard some original candidates, but never the other way around (that recurrences undetected in the scalar data set would show up afterwards). Intuitively, if (a) one generates a d-dimensional data set using whatever procedure, provided that one of the components of the new set is the original data and (b) the distance in d-

```
                                    main
 1   /*
 2        cretm.c Close returns
 3   */

 5   #define IMIN(a,b) ((a) < (b) ? a : b)
 6   #define MAX 16384                      /* number of data points */
 7   #define P 384
 8   #define ERR 32                         /* max average error */
 9   #define P0 64                          /* minimal period */
10   #define P1 2048                        /* maximal period */

12   #include <stdio.h>
13   #include <string.h>
14   #include <stdlib.h>
15   #include <math.h>

17   int main(int argc, char *argv[])
18 1 {
19 1    int i=0,j,p;                        /* i=initial, j=checkindex p=period */
20 1    int d1, delta, data[MAX];
21 1    int countper=0;
22 1    FILE *file,*per;
23 1    char name[20], nameper[20];

25 1    sscanf(argv[1],"%s",&name);         fprintf(stderr,"%s\n",name);
26 1    sscanf(argv[2],"%s",&nameper);      fprintf(stderr,"%s\n",nameper);

28 1    file=fopen(name,"r");
29 1    per=fopen(nameper,"w");
30 1    printf("Files opened\n");
31 1    while(EOF != fscanf(file,"%d",&d1) ){data[i++]=d1;}
32 1    fclose(file);
33 1    fprintf(stderr,"We read %d points on a %d-sized array \n",i,MAX);

35 1    for(i=0;i<MAX-P-P0+1;i++)
36 2      {
37 2        p=P0;
38 2        while(p<IMIN(P1,MAX+1-i-P))
39 3          {
40 3            delta=0;
41 3            for(j=0;j<P;j++) delta+=abs(data[i+j]-data[i+p+j]);
42 3            if (delta<ERR*P)
43 4              {
44 4                countper++;
45 4                fprintf(per,"%d %d err=%d\n",i,p,delta/P);
46 3              }
47 3            p++;
48 2          }
49 1      }

51 1    printf("There were %d candidates\n",countper);
52 1    fclose(per);
53   }
```

cr.c 1

Fig. 5.1   C.code for close returns

dimensions is the supremum of the Euclidean distances among each of the *d* components, then the claim is obvious. The operational rule-of-thumb, when it turns to actual data analysis, is: You may check for close returns in the original scalar data set. It is sufficiently good and much easier than waiting until after having generated a model dynamics (which we will do right away).

5.2.3   *Imbedding(s)*

The next step is to attempt a reconstruction of the data (and with it of the periodic orbits) in 3-space or on a Poincaré section. To this process, it is associated the concept of *imbedding* [1] that apart from its precise mathematical definition, came to encompass a number of data analysis procedures. In this sense, imbedding is an art. It entails choices and very little guidance is available. The comprehensive review [Abarbanel et al. 1993] appeared more than a decade ago is still the most adequate tool to start working.

The most general statement is that any C^1 manifold M of dimension m can be imbedded in Euclidean space $\mathbb{R}^n$ for n sufficiently large. With imbedding it is meant a smooth map $f : M \mapsto \mathbb{R}^n$ that (i) is a homeomorphism when restricted to its image $f(M) \subset \mathbb{R}^n$ and (ii) it maps the tangent space at any point $x \in M$ injectively onto the tangent space at the image point $f(x)$. Whitney's result [Whitney 1936] is that for any C^1 manifold M, imbedding is always possible given that $n \geq 2m + 1$. Of course, particularly nice manifolds may be imbedded in a space of smaller dimension (e.g., the graph of the linear function $z = x + y$ is an imbedding of $\mathbb{R}^2$ in $\mathbb{R}^3$), but for a general manifold the only safe statement is Whitney's. Consider as an example a closed non intersecting curve in 3-space in the form of an "almost" number eight (see figure). The curve is a 1-dimensional manifold while it "lives" in Euclidean space of dimension 3. Any attempt of describing such curve in $\mathbb{R}^2$ or $\mathbb{R}^1$ will fail because of some of the conditions above. Either there is no smooth mapping or it cannot be injective or it will fail to preserve distances or tangent vectors. Also, Whitney's statement says nothing about how f should specifically look like.

The fundamental work relating these ideas to data analysis has been initiated by Takens [Takens 1981]. Suppose that we have just a "scalar" time-series (a string of real numbers) but the collected data comes from a dynamical system of *a priori* unknown dimension. Putting it in more formal terms, we have some initial condition x_0, belonging to a manifold M of (*a priori* unknown) dimension m (moreover, the dimension of the

---

[1]The word goes back at least to Whitney [Whitney 1936], who calls *imbedding* the mapping of a manifold on $\mathbb{R}^d$. The word *embedding* is more frequently found in the physics literature, both for Whitney's and other uses.

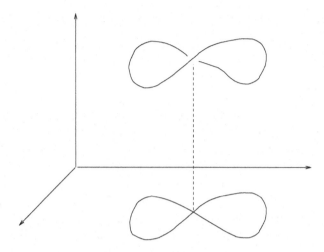

Fig. 5.2   Imbedded "number eight". A 2-D representation of the manifold on the horizontal plane is not an imbedding.

original dynamical system is at least m, but it may be larger[2]). Further, we have the time-evolution of this initial condition as given by the flow associated to the dynamical system, $\phi_t(x_0)$ and finally a smooth scalar measurement function $y : M \mapsto \mathbb{R}$. This material generates a time-series with measurements, $\{y(\phi_{t_k}(x_0))\}$, for $k = 1, \cdots, N$.

The practical imbedding task is now twofold: (a) Determine somehow the minimal dimension d such that the dynamics of $x_0 \in M$ is accurately described in $\mathbb{R}^d$, (b) Generate a d-dimensional time-series (of slightly shorter length at most $N - d + 1$) with the $\{y_k\}$'s that describes this dynamics. Both items are usually done in parallel.

We have to give some mathematical content to the expression "accurately described". The minimal dimension would be $d = 2m + 1$, only that we do not know m since we just have a scalar time-series. Even worse, we have to *assume* that the sampling was good and long enough to capture the features of the underlying manifold. Think of the number eight above. If we analyze a very short portion of that curve, we may miss the "pseudo crossing" point and regard the curve for all practical purposes as if it exists in $\mathbb{R}^2$ or $\mathbb{R}^1$. If our analysis turns out to give unexplainable contradictions later on, we may turn back to this assumption and criticize it.

A related question is, can we generate with the time-series a (sampled)

---

[2]Either the original x_0 was lying on an invariant manifold of smaller dimension than the whole system or some of the components were too small to be detected.

smooth orbit in d dimensions? Can this be done for many different d's so that we may afterwards choose an "optimal" value among those? What are the chances that (if we succeed) the orbit we obtain is an orbit of the underlying dynamical system except for at most a change of coordinates?

5.2.4   *Imbeddings and phase-space reconstruction*

There is a special case where the answer to these questions is easy. Consider a Hamiltonian System for a particle of mass one and one degree of freedom with a smooth enough potential function depending only on this degree of freedom. The Hamiltonian function coincides with the energy and is a constant of the motion: $H = E = p^2/2 + V(x)$. The dynamical equations read:

$$\frac{dx}{dt} = p \tag{5.2}$$

$$\frac{dp}{dt} = -\frac{dV}{dx} \tag{5.3}$$

Using a good numerical approximation method, from the time-series for x we can generate a time-series for the derivative $p = \frac{dx}{dt}$. We obtain thus a slightly shorter 2-D time-series with pairs (x, p), i.e., a time-series of the actual trajectory of the system in phase-space. In other words, from a scalar time-series in this case one can *reconstruct* the phase-space dynamics and with it pursue the dynamical analysis of the problem.

The generalization of this intuition would be that from a given scalar time-series of x we may generate time-series of discrete numerical approximations of the successive derivatives $x, \frac{dx}{dt}, \ldots, \frac{d^r x}{dt^r}$, with which to produce a $(r+1)$-dimensional time-series of a trajectory belonging to some $(r+1)$-dimensional dynamical system. Since taking numeric derivatives is a procedure that somehow enhances the intrinsic noise in the data [Mindlin et al. 1991], one may as well consider the string $x_k, x_{k+\tau}, \ldots, x_{k+r\cdot\tau}$ for some value of τ (this came to be called the *time-delay imbedding*[3]). For $\tau = 1$, this string is just a linear transformation of the crudest numerical approximation to the data and its first r derivatives. Let us illustrate it with the example of $r = 2$ (h is the sampling time-interval):

$$\begin{pmatrix} x(t) \\ x'(t) \\ x''(t) \end{pmatrix} = \frac{1}{h^2} \begin{pmatrix} 0 & 1 & 0 \\ 0 & -h/2 & h/2 \\ 1 & -2 & 1 \end{pmatrix} \begin{pmatrix} x(t-1) \\ x(t) \\ x(t+1) \end{pmatrix} + O(h) \tag{5.4}$$

---

[3]Just like this, with initial "e". τ or the associated sampling time is the *delay*.

One can also explore other values of τ and perhaps do better.

The result of Takens [Takens 1981] is that if the underlying manifold is a *compact*[4] invariant manifold of dimension m hosting the attractor of the system, then a $(2m + 1)$-dimensional (i.e., $r = 2m$) construction of this kind is generically[5] a good imbedding of the data, both for the time-delay approach or for the numerically approximated derivatives (perhaps the former is to be computationally preferred so that one avoids the introduction of ill-conditioning errors from h^{2m} for small h and large m, but just rescaling time so that $h = 1$ renders both approaches essentially equivalent in principle).

The intuitive version of this theorem is that, if the pair (ϕ, y) of chosen dynamical flow and measurement function is not an imbedding, then there exists a small modification of y that produces an imbedding. Care must be exercised, as usual, when using this intuition in practical situations. Takens' theorem does not say (contrary to what it is usually believed) that we have a "high probability" of picking a y that produces an imbedding. Also, small in mathematics means "as small as you need", while in natural sciences "small" is never smaller than the uncertainty of the data. Hence, Takens' theorem, in natural sciences, is a suggestion of what to try but it gives no guarantees.

In principle, not all dynamical systems can be rewritten in terms of $x, \frac{dx}{dt}, \ldots, \frac{d^r x}{dt^r}$. Such an approach will function only for systems responding to an equation of the type $x^{(r)} = g(x, x', \ldots, x^{(r-1)})$ *and* having such a nice function g that all underlying assumptions about dynamical systems still hold[6]. The suggestion inspired by Takens' results is that we may go ahead despite this observation, hoping that the reconstructed dynamics will be an imbedding of the original one.

It remains for us to pursue the analysis of the previous Chapters, collecting as much as we can of the topological information present in the data.

What on the other hand Takens' theorem does not say at all, is that the reconstructed dynamics differs from the original dynamics only in a

---

[4]Unlike Whitney's, Takens' results are formulated for compact manifolds, although Takens' imbedding theorem may be adapted to fit the non-compact case.

[5]A property is *generic*, according to Hirsch and Smale [Hirsh and Smale 1978] if it holds for a set $\mathcal{P}$ that contains a dense and open set.

[6]Try to do it with the Lorenz equations. The naive approach of modifying x, y, z as little as possible ends up in a rational function $g(x, x', x'')$ having x in the denominator. Hence, whenever an orbit crosses $x = 0$ (which is something that the Lorenz' attractor actually does) no meaningful conclusions are possible without further analysis.

smooth coordinate transformation in $\mathbb{R}^3$. We cannot conclude this just by using the theorem and whether it is true or not will have deep consequences for the rest of the analysis. Hence, we will attempt to retrieve topological information from the data, remembering that this information does not come purely from the data, but rather from the pair "data+imbedding" as long as there is no additional information about their internal relationship.

If one is prepared to admit the derivative of the data set as a possible imbedding coordinate, we may as well take the integral, or also some specific function involving both the data, its integral(s) and its derivative(s). In the same way, if one records the time series with a sampling interval h, we may perform the delay imbedding using some multiple of it, i.e., using $\tau > 1$. Let alone the fact that since we in general do not know m, we may want to produce many different imbeddings in different number of dimensions. On top of that we may want to pre-process the data in order to enhance its signal-to-noise ratio. So from one simple scalar data set there is a battery of procedures that can be applied and a battery of different imbeddings in different number of dimensions are possible. How are we to pick our straw in this forest?

5.3   Embedology

The title of this Section is just a fancy word coined by the practitioners of the art of imbedding [Sauer et al. 1991], in order to summarize the large amount of possible actions, tests, pre- and post-processing of the data that are available to the researcher. We repeat, Abarbanel's review [Abarbanel et al. 1993] is still among the most informative tools in this area. Our task can be summarized by the following actions:

(1) Record a scalar data set.
(2) Pick a not very large positive integer $D > 3$.
(3) For $d = 1, \ldots, D$ generate an imbedding in d-dimensions of the scalar data.
(4) Select the optimal imbedding dimension d_0 through one or more reasonability tests.

Furthermore, we have a "secret agenda" in this book called "let $d_0 \leq 3$", since the tools in the first Chapters of this book are useful only in 3-D. If $d_0 > 3$, a systematic plan of action is still missing (we will discuss this topic a little more in the next Chapter). So the second item in the list is quite easy. Since the secret agenda says $d_0 = 3$, let then $D = 4$ or $D = 5$ to begin with. The higher d's are there to verify if increasing the dimension beyond $d = 3$ yields some improvement in the reasonability of the data set. If not, $d_0 = 3$ (or less) and we proceed with our methods. If yes, we may increase D stepwise even further until finding the optimal dimension and then continue with some other analysis techniques. Still some partial characterization of the underlying system may be achieved using the tools discussed in this Chapter (find periodic orbits and other global properties of the data set).

The first item in our task-list is strongly dependent on measurement hardware. In practice, the theoretical scientist usually "obtains" a data set using the e-mail or the web, i.e., some experimental scientist, with or without a previous conversation with the theoretician, has already produced the best possible measurements available with his/her resources. We will not deal with this item but refer again to the references cited in [Abarbanel et al. 1993] in order to see some examples of data collection problems/features.

The tough part, of course, are the last two items. One may try different imbedding techniques in the first place.

(1) Time-delay imbedding ([Takens 1981]): $x_p(k) = y(k + (p - 1) * \tau)$, $p = 1, \ldots, d$. The d-dimensional array x runs over $k = 1, \ldots, N - p$ points. Recall that $\tau = 1$ means that one uses the sampling time-interval as delay factor, but one may want to use a delay-interval that is larger than the sampling time, which means letting $\tau > 1$. Which one to pick, it depends on the outcome of the "performance tests" (see below) for each imbedding alternative.

(2) Derivative imbedding (also [Takens 1981]): An example was shown previously. It can be seen as a linear rearrangement of the time-delay imbedding such that the outcome entries coincide with some numerical approximation of the successive derivatives of y. However this rearrangement involves a transformation matrix that could be ill-conditioned if the sampling interval h (see above) is very small and at the same time the order of derivation is very large.

(3) Integral imbedding ([Mindlin et al. 1991]): $x(k) = \sum_{j=1}^{k} y(j)e^{-\eta(k-j)}$. The (small and non-negative) parameter η is used to assure the numerical stability of the procedure, but it can again be established according to the performance tests. It is called "integral" since $x(k)$ is a crude numerical approximation of the integral $\int_0^k y(t)e^{-\eta(k-t)} \, dt$. The recursive computation of $x(k)$ is very simple. Define $y(0) = 0$ and then for

$k \geq 1$ we have that $x(k) = y(k) + y(k-1)e^{-\eta}$. In practice η is chosen so that $e^{-\eta}$ is slightly smaller than one.

(4) Transform imbedding ([Firle et al. 1996]): $x(k) = \mathcal{F}^{-1}[\sqrt{2\pi}i\mathcal{F}[y]](k)$, also called (discrete) *Hilbert* transform, where $\mathcal{F}$ is the discrete Fourier transform of the array y.

The list is far from complete. As a rule of thumb, integrals reduce measurement errors since you sum data points with errors that more or less "cancel each other" while derivatives enhance errors since you take differences and differences of differences.

5.3.1 *Strategies for choosing the delay-time*

Because the underlying dynamical system produces a smooth orbit, the points y_k and y_{k+1} may be very similar to each other for large portions of the recording if the sampling frequency is very large as compared with the typical oscillation frequency in the data set. On the other extreme, if those points are too far apart along the orbit, we will loose significant dynamical information. Some sort of trade-off may be necessary, and if there are no better reasons to decide in favour of some time-delay, one may want to choose some of the available criteria in the hope that consecutive recording points will bear significant and non-redundant information. A number of criteria had been advanced in [Abarbanel et al. 1993] such as to minimize the linear autocorrelation function (computed on the data as a function of the delay-time) or to minimize the average mutual information.

5.3.2 *Performance tests*

In this way we come to the final point, i.e., after having generated many different imbedding choices in many different dimensions, how are we supposed to pick the "optimal" one? The way to choose between different data-processing techniques is to identify the properties we will consider relevant and check how well our process satisfies those properties. The technique that performs best for the largest number of properties is the winner. If one has some previous information about the problem, then a test can be devised specifically for it, allowing us to be very restrictive. We list below some general consistency checks that may be useful in different circumstances.

5.3.2.1   *Distinct pseudo crossings ("inspection")*

There are just a few properties that a reconstructed phase-space dynamics must fulfill. Essentially, there is only one: The orbits are invariant sets and because of the unicity of solutions of differential equations, orbits do not cross, neither within themselves nor with other orbits. Apart from the very rare exact crossings, the next-best test is to inspect the data to see how portions of it relate to other portions.

In 3-D the inspection can even be done visually. Some portions of orbits that "cross" along the flow (actually, they do not cross as in a planar crossing, one portion goes above the other in the perpendicular direction) will later be involved in the computation of braids. Hence, it is important that these pseudo crossings are as separated as possible so that one may safely assume that e.g., portion "A" goes above portion "B".

5.3.2.2   *False neighbours*

Reflecting further about crossings, we may formalize the ideas on pseudo crossings a little more. In Figure 5.2, we can grasp the intuition behind the False Neighbours method [Abarbanel et al. 1993]. If the imbedding dimension is too low, points that are actually far away from each other in phase-space may appear to be very close to each other (perhaps because they are separated in the dimension $d+1$ that we have not yet computed). The worse "very close" situation is that two different points coincide (as above), which is a decisive rejection criterion for the imbedding. Coincidence in numerical procedures is a "rare" object, so let us fix ideas in the following way:

Definition 5.1   We say that two points are ϵ-*separated* in dimension d if their (Euclidean) distance, $\sqrt{(x_1 - y_1)^2 + \cdots + (x_d - y_d)^2}$ is larger than a given $\epsilon > 0$.

If we construct our successive imbeddings by adding sequentially dimensions one by one (the coordinates $x_1 \ldots x_k$ for the imbeddings in dimensions k and $(k+1)$ coincide), then two results are easy to get. If two points are separated in dimension k, they will still be separated in dimension $(k+1)$. Moreover, its mutual distance will never decrease when increasing the dimension, since we just add a non-negative term to the sum.

To establish a criterion here, is also a rule-of-thumb that the researcher adopts at her/his own risk. For a given fixed ϵ we may plot the number of non-separated points as a function of the dimension. If ϵ is too large, we run the risk of mistakenly consider pairs of points which are "there" for dynamical reasons as non-separated points. If it is too small, we run the opposite risk, a pair of points that may fail to coincide only because of roundoff errors may mistakenly be considered separated. The choice is ϵ sufficiently small (to allow for closeness intrinsic to the dynamics) but much larger than the typical roundoff errors (to prevent imbedding defects from remaining unseen). The optimal dimension d_0 is the smallest one yielding a satisfactorily small number (hopefully zero) of non-separated points.

5.3.2.3 *Singular value decomposition*

If the imbedding is performed for a dimension that is too large, we may expect that in some wise choice of coordinates the dynamical information will lie on a subspace of lower dimension, while the remaining components will only carry secondary (less relevant) information. Singular Value Decomposition [Aubry et al. 1991] (also known as Principal Component Analysis) has been proposed as a method of achieving the desired decomposition. Yet, in the present context it has also been criticized as highly misleading [Krmpotic and Mindlin 1997].

The idea is that out of the N points of imbedded data x in dimension d, we produce a $(d \times d)$ positive-definite symmetric matrix as:

$$A_{ij} = \sum_{k=1}^{N} x_{ik} x_{kj}.$$

Hopefully some eigenvalues of this matrix will be considerably large, while others will be around zero, of about the size of the expected errors in the data. We may hence see a clear gap in the spectrum, d_0 eigenvalues lie above the gap and $d - d_0$ are below. If this is the case, we pick the eigenvectors associated to the eigenvalues above the gap as the new coordinates, and declare our optimal imbedding dimension to be d_0.

5.3.2.4 *Fractal dimension*

Sauer, Yorke and Casdagli's version of the imbedding theorem [Sauer et al. 1991] modifies Whitney's and Takens' results in the following ways: (a) Instead of Takens' compact manifold of dimension m it uses an underlying compact invariant set A (think of "the attractor"), (b) instead of the proper dimension m, it uses an estimate of the fractal dimension (specifically, the so-called box-count dimension[7]) of A, and (c) instead of "generic" it uses the concept of *prevalence* [Sauer et al. 1991] that roughly speaking can be understood as "probability one". Intuitively, since the imbedding dimension has to be at least $d = 2m + 1$, we may restate it as $d > 2m$ and integer. Sauer's claim is that now we can use the box-count dimension of the attractor instead of m.

In this way, if we have some hint about this dimension, we may establish some lower bound to d. In particular, we may again imbed the data in many different d's, compute the fractal dimension of the imbedded data with some satisfactory method and check the outcome. If the value of the fractal dimension essentially stops growing (it becomes approximately constant) when going above some (manifold) dimension d_0, then we take d_0 as the optimal imbedding dimension.

5.3.2.5 *Surrogate data*

Whatever test we apply to our data, it should be contrasted against something. The usual statistical procedure is to have a null-hypothesis that may read more or less as follows: The *data set* cannot be distinguished using the test X (also called the *discriminant statistics* from other data sets produced with the *alternative random method*. If the outcome of the discriminant statistics is significantly different for our data set as compared with a large family of other strings of numbers produced with the alternative random method then we can say that our data is unlikely (with a given probability) to be a member of the random family.

The method of *surrogate data* [Abarbanel et al. 1993] begins then by constructing one or many new data set(s) (the surrogate) where the property we are considering is destroyed on purpose. Then we proceed with any of the tests above both for the real data set and the surrogate(s) and check if the differences in the outcome of the discriminant statistics are significant. Then, in a weak inductivist manner, we are allowed to believe that our discriminant statistics is sensitive to the selected property.

An illustration of what we can expect using this methodology was given above when considering close returns. Certainly, solutions of ODE's are smooth and deterministic. Sampled data taken from the Lorenz system has these characteristics built-in. When we randomly scramble the data

---

[7]Again, Abarbanel's review [Abarbanel et al. 1993] is a good source to learn how to compute different kind of dimensions for a data set. You may even check [Solari et al. 1996a] for this purpose.

set collected from a numerical integration of the Lorenz system, in most of the cases the smoothness and the determinism are destroyed. The new (surrogate) data set contains data values that belong to the Lorenz system, but all the dynamical information (given by the time-ordering of the data set) is no longer there. The discriminant statistics *find close returns* will then yield dramatically different results in both cases. If we were given the data set without disclosing its origin and also the surrogate data, we could have concluded, comparing the number of periodic orbit candidates, that the data was unlikely to belong to the set of surrogates. Other examples of surrogate data analysis can be found in [Solari et al. 1996a].

5.4 Reconstruction of the Poincaré Map

The scheme suggested so far in this Chapter is more or less the following: (a) collect data, (b) find close returns, (c) reconstruct phase-space dynamics, (d) verify that the dynamics fits in 3-D, (e) find a Poincaré section, (f) read the braids associated to the periodic orbits. In this situation, ultimately, one can derive a reconstructed Poincaré map using the reconstructed dynamics and the associated Poincaré section.

There exists experimental data that do not fit the previous phase-space based programme very well. In particular, (c) fails[8]. For example, in a pulsed laser, it is expected that the signal will grow and decay sharply in a pulse-like way, being essentially zero in the time-interval between pulses. The available information in the collected data amounts to the shape of the pulses as well as the "dead" interval between pulses.

Any attempt to generate an imbedding using e.g., time-delay, will result in very high dimensions. In fact, as long as the imbedding dimension is smaller than the typical number of data points in the dead intervals, the imbedded data will present self-intersections (exact ones if the data has exactly zero signal in the dead interval). To fix ideas, if the dead interval after two different pulses consists of $Q > 1$ data-points, an imbedding with $1 \leq R \leq Q$ data points will generate two different orbits (those passing through the two different pulses), both of which eventually pass the R-dimensional point $(0, \ldots, 0)$.

In order to be able to distinguish both situations, one needs more than Q dimensions. In fact, let $s_1, \ldots, s_k$ and $r_1, \ldots, r_k$ represent the last k points in each pulse (which are satisfactorily different). Then the points

---

[8]We will see later that (d) may fail in a non-trivial way.

$(s_1, \ldots, s_k, 0, \ldots, 0)$ and $(r_1, \ldots, r_k, 0, \ldots, 0)$ in $Q + k$ dimensions, corresponding to portions of imbedded orbits in phase space, are different. Paradoxically, the better the data collection, the larger the dimension.

Let us compare this situation with population data from e.g., birth of sea-elephants along the year. All births in a sea-elephant colony occur concentrated in a relatively short time-interval in spring. The colony swims to a suitable land spot, birth takes place and after a few months the colony swims in open sea again. Birth records will consist of pulses, having a maximum in spring followed by a more than half-year long period with zero births. A similar thing can be said for time-records of egg-laying by many seasonal insects.

The fact that potentially simple dynamics demands a large imbedding dimension may be a conflict arising from the choice of methodology. In any case, if our data has large dead times, it is relevant to consider if an ODE dynamical system is actually the best tool to describe it. In order to understand the dynamical features, it is perhaps enough with a few variables characterizing the pulse and the dead time[9]. Following the biological intuition, it might be wiser to attempt a direct reconstruction of the Poincaré map without passing through the phase-space reconstruction, especially in these cases where resorting to phase-space dynamics is an obstacle rather than a helpful tool.

5.4.1   *Sampling the Poincaré map*

An example of this new situation is the case of laser dynamics described in [Solari et al. 1996b]. We will describe here a general methodology to deal with such problems.

In Figure 5.3 we show some typical data.

The underlying experiment is a laser with saturable absorber [Fioretti et al. 1993]. It consists of a Fabry-Perot laser cavity containing an absorbing cell. The absorbing properties of the molecular gas placed in the cell *saturate*, i.e., the gas stops absorbing beyond a certain threshold level[10].

Inspection of the figure suggests a way of action in order to characterize the pulse. First one may decide a threshold L under which the data can be assumed to be zero. This may lie around $L = -195$ in the picture.

---

[9]This depends, of course, of the degree of detail in the description that we deem necessary.

[10]For comparison with present-day experimental resources, we note that this experiment was recorded with 8-bits resolution [Fioretti et al. 1993].

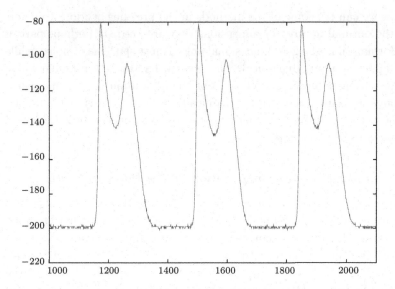

Fig. 5.3   Pulse-data from a Laser with saturable absorber (see text). The horizontal axis displays the position of the plotted points in the collected array. Successive points were recorded with a time-interval of $200ns$. The vertical axis yields the outcome of the measuring and digitalizing device, in arbitrary units.

Another threshold M can be defined to identify the "tip" of each pulse, it might be e.g., $M = -100$ (or $M = -120$ if one wants to get also the lower tip). Consecutive data points lying above L can be used to characterize the pulse, together with an entry for the pulse duration (the number of points between two consecutive passes from below L to above L) and perhaps another entry for the tip of the pulse. The actual maximum is likely to lie between two consecutive recordings, if one uses even sampling without any special feature adapted to detect maxima. Hence, we might want to interpolate consecutive data points above M (with Lagrange interpolation, splines or some other suitable interpolation technique) and compute the interpolated maximum as a better estimate of the tip value and position. In this way, each pulse is described by an array with about 200 dimensions, as can be seen from the picture. We can regard successive pulses as the image of each other as given by the Poincaré map.

So far, little has been gained as compared with modeling phase-space with a couple of hundred dimensions (which is more or less the size, in data points, of the dead periods). The overall gain is that we eliminate an intermediate step (reconstructing the flow) and with it a source of error.

Now we can try some of the methods in the previous Sections to determine the optimal number of dimensions that are needed. Perhaps part of the information in the 200-dimensional pulse is just spurious or noise, while a small part is mainly dynamical. The guess at hand here is Singular Value Decomposition as in Section 5.3.2.3. For the data in question it happened that only two dimensions were dynamically relevant [Solari et al. 1996b], and the machinery of associating braids to periodic orbit candidates could be applied as in Chapter 4.

5.4.2   *Finding the Markov partition on the Poincaré section*

Having a considerable amount of information about a system raises the temptation of establishing finer details of the dynamics rather than stopping at a coarser description. For example, if we know in advance that our data comes from the Lorenz equations, or from a suspension of Smale's horseshoe, we may wish to extract from the data specific details of the Lorenz or Horseshoe attractor as further support to our characterization of the data.

Focusing on the title of the Section, we may want to identify from our reconstructed phase space (or Poincaré section) the 0 and 1 strips of the horseshoe, or which half of the Lorenz' section the data is crossing (around which of the two unstable fixed points the orbit is circulating at a given moment). The "hidden assumptions" are then numerous:

(1) The underlying chaotic invariant set is known or at least the associated template is fully identified.
(2) A large number of periodic orbits from the invariant set are known.
(3) The information on each orbit is enough to unambiguously identify its braid inscribed on the template (braid and braid word, i.e., which strand lies on which branch of the template).
(4) The periodic orbits cover the Poincaré section sufficiently tight (the distance between any two points of the data set is smaller than some threshold $\delta > 0$ when necessary).

When all these assumptions are satisfied, we may attempt the following procedure: (a) Divide the Poincaré section in small cells (circles) of radius $\epsilon > 0$. (b) Choose the cells in such a way that at most few elements of the set of periodic points on the section (corresponding to the periodic orbits of the flow) lies on each cell. (c) Assign to each cell the symbolic name of the periodic point(s) lying on it (or nothing if the cell was empty). If there are many points in one cell having *different* symbolic names, this just means that the size of that cell is too large; a finer partition is needed there.

If we are lucky and nothing went wrong, we may end up with some connected regions on the Poincaré section sharing the same symbolic name. If the symbolic alphabet only had two letters and no cell is empty, we may in addition generate a borderline between the 0 and 1 regions having thus partitioned phase space according to the underlying Markov partition. A refined version of this intuitive procedure has been computationally implemented on [Plumecoq and Lefranc 2000a;b] for a numeric data set obtained through integration of a system presenting a Smale horseshoe as underlying attractor. Some thousands of orbits were necessary to achieve a satisfactory description. Note that for experimental data the typical amount of reconstructed orbits lies in the region 10 – 100.

5.5   Occam's Razor

This Chapter has dealt with methods of collecting and interpreting experimental data. The precarious terms in which the discussion is presented is intrinsic to the problem, since experimental data is always bound to uncertainties, collecting errors, measurement errors and roundoff errors. The dynamical information is mixed up with this error and a subsequent separation is neither easy nor complete. One has to test different tools and make decisions regarding how well the tests are passed (i.e., pick up a threshold ϵ for the gap in the spectrum of A or for the separation among data points). The degree of confidence with which we declare the tests passed puts a limit to the degree of confidence we can assign to our subsequent analysis. Furthermore, some of the criteria are not deductive but actually inductive, i.e., they result in reasonable conjectures that we accept provisionally in order to proceed further with the study. Incorrect conjectures may show up later on as contradictions or other types of difficulties.

Decision making is guided by a principle attributed to the medieval friar William of Ockham, stating that the explanation of any phenomenon should make as few assumptions as possible, eliminating, or "shaving off", those that make no difference in the observable predictions of the explanatory hypothesis or theory. When given two equally valid explanations for a phenomenon, one should embrace the less complicated formulation (see the article in Wikipedia: `en.wikipedia.org/wiki/Occam's_Razor` for further information). The principle seems to have been invoked in science for the first time by Hamilton in the 19th century. This idea gives some structure to the fact that some degree of uncertainty is unavoidable when analyzing experimental data, since "unclear choices" and "less complicated formulations" always carry along some degree of subjectivity.

The subject of "simplicity" has been addressed in the epistemology of the natural sciences [Popper 1959]. Popper, elaborating over a criterion introduced by Weyl, puts the criterion in terms of falsability and "empirical content" of a theory [11].

5.6   Final Remarks

In this Chapter we have discussed methods to analyze data sets in phase-space, i.e., in the flow associated to some ODE-dynamical system. For the case of major interest in this book, i.e., 3-D dynamical systems admitting a Poincaré section, there may be another choice, namely that of reconstructing the Poincaré first return map directly instead of first reconstructing the flow and subsequently computing the resulting model for the Poincaré map. We defer the discussion of this alternative to the next Chapter.

Also the fine-tuning of data analysis, caveats, problems and intrinsic limitations will be considered in more detail in the coming Chapters.

---

[11]Comparing two theories, one is simpler than the other when it gives more opportunities (tests) of being falsified.

Chapter 6

Reconstruction of Phase-space Dynamics – Advanced Course

6.1 Introduction

This is the advanced course in data analysis. We consider here situations where the methods of the previous Chapter are insufficient, or not optimal, or ambiguous. We will produce and discuss some methods and techniques which are not always supported by Theorems or explicit rules, but which have shown to be useful in particular examples, or at least indicate an interesting way to pursue research. The choice of discussion topics is arbitrary, only guided by what the authors know (and do not know) as well as what they consider interesting. No claim of completeness is done.

6.1.1 *Epistemological ruminations*

As soon as experimental data entered the picture, we moved from mathematics into theoretical natural sciences (say theoretical physics if you like it). Now we have to deal with the fact of our intrinsically incomplete knowledge of the problem. We want to be particularly careful in this section regarding our point of view on the scientific matters involved in the discussion, making them explicit, since part of the discussion surrounding this subject conceals the fact that one may be using different epistemological systems.

Our data is a partial and particular probing of nature that is used in two different forms in our search for deeper understanding, namely to generate an explanation and to test this explanation. A theory, in order to be considered as scientific it has not only to explain the observations but also to produce predictions that are testable [Popper 1959]. In a first step, data helps us to produce explanatory hypotheses such as "the orbits found are organized in the same form as the orbits in a horseshoe". Such hypotheses are subsequently incorporated in a "theory" (elaborated by induction) e.g., that all the orbits of the system are organized as in a horseshoe, or the more complex theory (in Popper terms as advanced in the previous Chapter) stating that all the orbits of the system *imbedded in the proposed way* are organized as in a horseshoe. The second step rests in the concept of falsability [Popper 1959, Lakatos 1978], in the sense that further collection of data may prove the theory wrong, or even the same data set may prove the theory wrong if a prediction deduced from the theory turns out to be incompatible with the data. This is as much as we can expect from a theory in natural sciences: It holds as long as it is consistent and not falsified. We can also make our theory still more difficult to falsify (more complex and less falsable in Popper's sense), i.e., with lower empirical content. For example, if we restrict our predictions to those that are independent of the imbedding.

The two main theoretical frameworks that are relevant for the purpose of this book are knots and braids. After using the procedures of the previous Chapter, one may attempt to identify periodic orbits as knots or braids and proceed with the construction of a theory. Since several different (inequivalent) braids can be associated to the same knot, it is clear that comparing experimental information based upon the knot-content produces a more complex theory than one based upon the braid-content. Every time the knot theory is found making a wrong prediction, i.e., being false, then the associated braid theory is also wrong. The reciprocal is not true, since different braids may correspond to the same knot, and hence braid-predictions may be incorrect but the corresponding knot-prediction still be correct.

The trade-off between complexity and risks taken in the predictions is clear. Once again, it is a matter that the scientist must evaluate and decide based upon her/his convictions in the particular case studied. For example, if we are to infer a template from data after some successful imbedding, it is possible to eliminate almost all the risk by placing each reconstructed strand in an associated strip. Such template will have no predictive power and as such it will not be a scientific theory since it is merely descriptive and not falsable. Hence risk, seems to be unavoidable.

What can we predict? There are several type of predictions that can be made (ordered from larger to smaller empirical content):

(1) Spectra of orbits. Which orbits are implied (forced) by the orbits found. This step can be done so far in terms of braids only and requires to

trim the template in the manner explained in Chapter 4.

(2) Topological organization. Induce a template and sustain the theory that all the orbits eventually found can be drawn with the induced template producing the correct braid or knot type (strong -simple- and weak -complex- version, respectively).

(3) Topological organization up to template differences. This is, either by prescribing a chosen imbedding or by using imbedding-invariant characteristics [Ghrist et al. 1997].

An additional problem appears when the theoretical apparatus introduced for periodically forced flows in $\mathbb{R}^2 \times \mathbb{S}^1$ is extended to autonomous flows in $\mathbb{R}^3$. A global Poincaré section may not exist and the use of braids is then less natural than in periodically forced flows.

6.2 Templates, Braids and Braid Words

There exists a hierarchical relation among periodic orbits of flows, imbeddings, braids and templates.

The basic object in our approach has always been the experimental data set. On top of that we produce an imbedding and recognize periodic orbit candidates. These are the most fundamental objects and their quality is decisive for the accuracy of our future predictions.

Given the periodic orbit and the imbedding (be it on the 3-D flow or on its 2-D Poincaré section) we can compute the braid (or rather the braid type) unambiguously. This is the "next most basic" object and it is enough for computing e.g., linking numbers as in Chapter 2 or the minimal periodic orbit structure and topological entropy estimates as in Chapter 4.

The template and the "name" of the braid in terms of a symbolic alphabet are less basic objects. This may be illustrated by the following facts, discussed already in Chapter 2:

(1) The same braid type may have different names *within the same template*. This was illustrated already in Chapter 2 where we comment in the discussion around Figure 2.10 that the same braid corresponds to two different periodic orbits of the horseshoe (with different symbolic names) and also in Chapter 4, where we discuss the ambiguity induced by the fold in naming periodic orbits.

Taking again the horseshoe as an example (where most features can be computed exactly), a given braid type may be associated to many different braid words, beyond the ambiguity induced by the fold, just because of the conjugation equivalence relation within a braid type. For example, the braids labeled 8_5 with braid word 00101011 and the braid 8_6 with word 00111011 in [Mindlin et al. 1993] correspond to the same braid type. Many other examples are displayed and organized in [de Carvalho and Hall 2003], see Figure 2.11. In fact, a great deal of structure concerning conjugacies and forcing order can be established among classes of horseshoe braids.

(2) The same braid type may have different names *on different templates*. This statement is perhaps less surprising, since different templates may have a different number of branches and hence different symbolic alphabets. A braid fitting two different templates could have a symbolic name with e.g., two letters in one template and with three letters in another.

6.3   Knots vs Braids: Freedom of Choice of Poincaré Section

Imbedding knot holders in $\mathbb{S}^3$ gives rise to a surprising richness. A theorem in [Ghrist et al. 1997, page 106] indicates that *"Any orientable template may be imbedded in $\mathbb{S}^3$ so as to contain an isotopic copy of all orientable templates as disjoint separable sub-templates"*. There are also templates called *universal* that contain all possible knots. Further, it is suggested that the templates related by different imbeddings should be compared using template-invariants.

The important question here is how are we going to compare templates in terms of data analysis and the identification of a finite set of periodic orbits. Should we check that two templates yield the same knots or should we check they yield the same braid types? Picking one or the other criterion gives dramatically different results.

Historically, knots appeared before braids concerning its applications in data analysis. Indeed, the whole template programme had knots in mind from the beginning. The question *"Does a template host all possible knots?"* (i.e., is it universal?) has a simpler answer than the corresponding question for braids, since there are in some sense "less" knots than braids, meaning that recasting a periodic orbit either as a braid or as a knot, the resulting classification of orbits is different. Many different braids (or braid-types) are associated to the same knot.

The question about universal templates is answered in the previously mentioned book of Ghrist, Holmes and Sullivan [Ghrist et al. 1997]. There exist universal templates, their features can be identified and moreover any orientable template can be recast as *universal* (i.e., containing copies of all possible orientable templates and hence of all possible knots hosted by them).

However, the method of proof used in [Ghrist et al. 1997] to establish these results is highly knot-dependent. If one can recast the periodic orbits present in a template as braids, i.e., in a situation where e.g., a global Poincaré section valid for the whole template exists and when this section is a sufficiently simple surface (a topological disc), then an extra constraint is automatically imposed, namely that of the period, counted as the (integer) number of times the orbit visits the Poincaré control section. The period reveals itself in the braid as the number of strands, which is also a topological invariant of the orbit. If applied to braids instead of to knots, some of the proofs in [Ghrist et al. 1997], would alter the number of strands, or equivalently alter the nature of the Poincaré control section by e.g., redefining it as a (not necessarily connected) subset of the original one. Eliminating some regions away from the control section, one or more passes through the section have to be recast as something different. The natural clock that the control section gave to the system now misses some "ticks" every now and then, when the system visits some special portions of phase-space.

Let us illustrate the question with a *gedanken experiment*. Imagine a pulse-laser as the one presented in Section 5.2.4. Each pulse is a natural candidate for being a point on the control section. Pulses are separated by rather long zero-intensity periods and can thus be clearly identified[1]. The analysis in [Solari et al. 1996b] proceeded along these lines. Now we could claim that this laser was not a pulse-laser but a *two-pulse-laser*, i.e., that we have a pass through the control section only every second pulse. Without pursuing the imaginary experiment further, we claim that such an analysis would yield internally consistent results, but different from the previous analysis. Moreover, by dropping the first half-pulse we can produce two different two-pulse descriptions of the same data.

Which picture is the correct one? For this far-fetched example, one

---

[1]If you really want to complicate things further, you may notice that each pulse in Figure 5.3 has two tips, so one may want to count them as two different pulses. However, in order to do this one should find more experimental support. At a first glance it sounds artificial, since there is no zero-intensity period between the tips and hence the "pulses" do not behave as such, i.e., as a bunch of energy that is expelled in one shot, subsequently emptying the energy content of the system.

may invoke *Occam's razor*. With the available data, the assumption that the laser produces double pulses (or, correspondingly, half-pulses) is not the simplest possible one supported by the observations. It requires to assume arbitrarily and without favourable experimental support that the zero-intensity periods belong to two different classes: inter-pulse zeroes or intra-pulse zeroes[2]. But in fact there is no observable difference between the even zero-periods and the odd zero-periods. At least no difference that could be noticed by the researchers producing and analyzing the data. Even if the assumption is not clearly incompatible with the data, it is so far unsupported by a discriminant test. Hence, the description without such an assumption is simpler (in Weyl and Popper's sense [Popper 1959]) and it has all its (remaining) assumptions exposed to discriminant tests.

If one wants to pursue the double-pulse line anyway, it remains to establish by further analysis of the data if the assumption leads to predictions that are distinguishable from those arising from the single-pulse assumption. If this is the case, then a discriminant test should be performed in order to establish which of both pictures better fits the experimental situation.

The bottom-line for the researcher is to focus on the following questions: (a) How reliable is my choice of Poincaré section? (b) Are there alternative (inequivalent) choices of Poincaré section? (c) What are the reasons I present for choosing among alternatives?

6.4   Topologically Inequivalent Imbeddings

Let us consider further the question of the choice of Poincaré section. In any numerical reconstruction of an attractor we have only a finite number of data points, i.e., most of the phase space is **not sampled**. In numerical experiments one can usually refine the sampling just by taking more and more points in the regions of interest (within certain limits). The situation is usually worse with experimental data.

The "holes" where we lack information are almost everywhere, yet we usually (implicitly) assume that there is a smooth interpolation of the data we have, and that the interpolation has no singularities (Occam's razor again; this is the simplest assumption). For example, we always assume that in the unsampled holes there are no regions where the orbits intersect.

---

[2]Unless all zero-periods are exactly identical (whatever this means), there is always a tiny hope that they actually could be different.

Also, in the "tightly sampled region", we use non-intersection as a quality control. We reject reconstructions leading to orbit intersections.

Some questions arise naturally: (a) To what extent these hypotheses are satisfied? (b) Does the topological information recovered by the reconstruction depend on the specific features of our imbedding procedure? (c) What are the consequences (implications, predictions) of the reconstruction?

In an exploration of these questions delay imbeddings were considered [Mindlin and Solari 1995] and the braid structure was reconstructed for delay-times in a wide range. The imbedded data turns out to be compatible with having a large disc as Poincaré section, where the data falls tightly on three "islands" of intersections with the Poincaré section[3], separated by large regions with no available data (the "hole").

Regarding the whole imbedding procedure as a function of the chosen delay-time, two disjoint delay-time intervals were found where the imbeddings yielded acceptable reconstructions. These intervals were separated by an interval of delay-times where the reconstructed flow presented apparent self-intersections. Remarkably, the braid organization of the orbits was *different* in each of the two time-delay intervals yielding acceptable imbeddings.

In this specific example we may connect these observations with the considerations of the previous Section. One possibility is to extend the definition of Poincaré section, and then select a different (extended) Poincaré section [Tsankov et al. 2005]. The "old" periodic orbits of period $3n$ are now recast as periodic orbits of period n by considering the braids associated only to strands with initial points belonging to one (chosen) island. This alternative definition of the Poincaré section automatically eliminates all topological ambiguities for the case $n = 1$, although the situation may not always be so simple. The old periodic orbits of period 3, having two different topological structures for different imbeddings, become now period-1 orbits (of which there is only one braid type). We observe that, restricting the region of the phase space where predictions are made, the empirical content of the theory is reduced as well as its falsability.

Of course, one should not stop at period 3. What happens with the orbits of (old) period 6 and 9 (that now become of period 2 and 3), are there other orbits whose period is not a multiple of 3? What happens with the global torsion around the new period-1 orbits? Are these objects different or not when the delay-time is changed? In any case: is there

---

[3]Recall, however, that the choice of Poincaré section is not necessarily unique.

experimental support (or some other satisfactory enough and scientifically testable support) to prefer one choice of Poincaré section to the other? The situation is still far from being completely understood.

Regardless of the specific positive or negative answer for this particular problem, the reader may realize that in the general case, the unexpected ambiguity arising by one "innocent" choice of Poincaré section need not disappear simply by choosing another section. Different sections may give rise to different ambiguities. It might happen that some lucky problem has a lucky choice of Poincaré section that is free from ambiguities of any kind, but there is no reason to believe that this will always be the case if no additional information is provided.

The bottom line, again, is that one cannot use imbedding techniques as "black-boxes". There are a number of choices to be done. Some of the choices we may say, are unexpected. Each choice has to find support in the available experimental (or other) setup and this support should be exposed to criticism, i.e., one should state as clearly as possible, the reasons for making a given choice, in such a way that these reasons can be tested against alternative choices. The consequences of each choice should be analyzed as deeply as possible, since this itself may constitute a good decision test.

6.5   Do Imbedding Techniques Influence the Resulting Topological Invariants?

The fact that different types of topological fine-structure can be produced with the same data suggests that the fine details of the braid structure might depend on our reconstruction, since they in fact depend on the interpolation performed. We will further consider in this Section whether this dependence is only a matter of, say, insufficient sampling or if it hides something more fundamental. This question has a number of different features to be considered.

6.5.1   *Imbeddings and reconstruction of the dynamics*

To imbed a time-series and to reconstruct the dynamics that generated the time-series are two different things that are often confused in the literature. The reason for the confusion probably lies in the intuitive presentation of the previous Chapter, namely that both methods have a great deal in common for certain classes of dynamical systems.

The imbedding ideas behind Whitney's and Takens' theorems aim to map an M-dimensional manifold to $\mathbb{R}^N$, where $N = 2M + 1$ in such a way that prescribed properties of the original manifold are preserved by the map. To fix ideas think of a time-series that exactly records a portion of an orbit from a 3-D dynamical system (i.e., a 1-dimensional manifold), or even a time-series that approximately records an attractor of low box-count dimension (below 1.5). Then Whitney's theorem (or Sauer's version) assures that the manifold underlying the record can be mapped on $\mathbb{R}^3$.

Takens' theorem further says that either a delay (or derivative) imbedding or some other imbedding infinitesimally close to it will do the job. This is both good and bad. It is good that some imbedding exists and that it looks close to a procedure that is familiar to us. It is less good that no guarantees are given. "Infinitesimally close" is a mathematical concept but it is foreign to natural sciences. One may try and try for centuries and never hit the proper imbedding (that lies infinitesimally close to our trials).

We insist in that these theorems do not say and cannot say what is the connection between the imbedded 3-D dynamical system and the original 3-D dynamical system generating the data. We will present some results stating that for certain special cases the job is performed as good as possible by a special choice of Takens' procedure.

6.5.1.1   *A theorem on periodically forced oscillators*

Periodically forced oscillators can be described in the following way:

$$\begin{aligned}
\dot{x} &= y \\
\dot{y} &= f(x, y) + A \cos(\omega\theta) \\
\dot{\theta} &= 1.
\end{aligned} \tag{6.1}$$

where $f(x, y)$ represents the oscillator force (for a harmonic oscillator we just have $f(x, y) = -kx/m$), and A is the amplitude of the periodic forcing. The variable θ is a surrogate for time, rendering the original non-autonomous forced system into an autonomous one. Since the forcing is periodic, $\theta \in \mathbb{S}^1$ and any fixed $\theta = \theta_0$ is a good Poincaré section. The clock-time between two consecutive passes through the Poincaré section is $T_0 = 2\pi/\omega$. For these systems, we will review and develop some ideas advanced long ago [McRobie and Thompson 1993].

Lemma 6.1   *Periodic orbits in forced oscillators such as 6.1, having Lipschitz right-hand side, yield positive braids.*

Proof.   A periodic orbit of the system projects as a smooth closed curve on the plane (x, y). An initial condition on such curve makes one or more complete clockwise revolutions after a clock-time $T = nT_0$, where $n \geq 1$ is an integer. The revolution is forced to be clockwise since positive y forces x to increase (it moves to the right) and negative y forces it to decrease. We call such an orbit a *period-n* orbit[4]. The orbit has n different intersections with the control plane $\theta = \theta_0$. Each arc along the projection of the periodic orbit on the (x, y)-plane joining consecutive intersection points (consecutive along the curve) constitutes one strand of the braid describing the orbit. Two arcs on different strands involved in a strand crossing will have the same x-coordinate at the crossing point (same value of $t \in \mathbb{S}^1$ as well). The arcs at that point are ordered one above the other according to the value of the y-coordinate, namely $\dot{x}$. Circulating according to time-orientation, hence, all strand crossings are *left-over-right*, since the strand evolving in time from small x-values to large x-values has larger derivative (hence y-value) than the strand evolving in the opposite x-direction. Therefore, all strand crossings are left-over-right and the associated braid is positive. $\square$

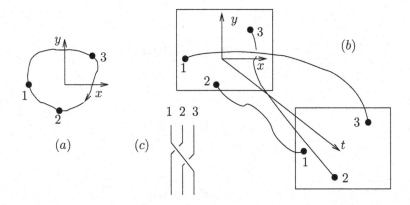

Fig. 6.1   Illustration of McRobie's Theorem: All braids in periodically forced oscillators are positive. (a) A projection of a period-3 orbit on the (x, y)-plane indicating the three intersection points. (b) The orbit in (x, y, θ)-space evolving between two copies of the Poincaré section, $\theta = \theta_0$ and $\theta = \theta0 + 2\pi$. (c) The schematized projection of the orbit on the (x, θ)-plane shown as a positive braid.

---

[4]We note on passing that not any closed curve in the $(x, y)(t)$-plane is a possible periodic orbit of the system. The additional constraint of y being the derivative of x carries along a restriction on the Jordan curves that can be associated with periodic orbits of the system.

We illustrate the result in Figure 6.1. Consider to fix ideas the leftmost intersection in the x coordinate as initial condition, and label the intersection points sequentially in increasing order according to their x-coordinate (not necessarily the same order as the visiting order along the parameterization in time).

Corollary 6.1 *If a data set can be properly imbedded in the coordinate system $(x, \dot{x}, \ddot{x})$ and the plane $\ddot{x} = c$ (where c is a constant) is a Poincaré control section, then the derivative imbedding yields positive braids.*

Proof. The situation can be transcribed to the previous problem since the projection of periodic orbits on the $(x, \dot{x})$-plane behaves exactly as above. □

Corollary 6.2 *If a data set can be properly imbedded with a 3-D delay imbedding $(x_1(t) = x(t), x_2(t) = x(t + h), x_3(t) = x(t + 2h))$, with delay-time h sufficiently small, there exists a plane in (x_1, x_2, x_3)-space which is a Poincaré control section equivalent to $\ddot{x}$, and the delay imbedding yields positive braids relative to this Poincaré section.*

Proof. If h is sufficiently small, there is a linear transformation up to order $O(h)$ between the delay coordinates and the derivative imbedding (see Eq. (5.4)). Then, the argument of the previous corollary holds as well on the control plane corresponding (up to $O(h)$) to $\ddot{x} = c$. □

The lesson from these results can be summarized as follows:

(1) Nice enough periodically forced oscillators can be described with positive braids.
(2) Derivative imbeddings and time-delay imbeddings with small delay can be described with positive braids, in the favourable case that the plane given by $\ddot{x} = c$ is a Poincaré control section.
(3) The imbedding coordinate system given by derivative imbeddings or time-delay imbeddings with small delay in the above situation is a natural coordinate system to describe periodically forced oscillators.

These facts explain why in simulations and experimental analysis positive braids appear with extraordinarily large frequency.

6.5.2 *Imbedding as a coordinate transformation*

Lured by the previous results, it has been the belief/hope of many a scientist that the imbedding coordinates would be, if not the natural coordinate system to describe an experiment, at least a good coordinate transformation from the (possibly unknown) original dynamical system.

The mathematical formulation of this hope is that

Proposition 6.1   *(i) The data comes from a dynamical system, ODE, in* $\mathbb{R}^2 \times \mathbb{S}^1$.
(ii) There exists a coordinate transformation from the original ODE to the time-delay-coordinates.

While this proposition may hold in some cases for forced oscillators, the results of [Mindlin and Solari 1995] allow us to conclude that it is not true in the general case of imbedding of a scalar data set. Indeed, whenever two imbeddings yield topologically inequivalent braids, at least one of the imbeddings cannot be regarded as a homotopy from the original dynamical system since there are no coordinate transformations in $\mathbb{R}^2 \times \mathbb{S}^1$ mapping a braid onto a topologically inequivalent braid. If we do not have additional information from the original dynamical system, we do not even know which of them is not a coordinate transformation!

At best, we can safely state that the topological properties computed by imbedding a scalar data set in 3-space contains information arising from both the original system and the imbedding procedure. To distinguish which part of the information comes from each ingredient is in general impossible without additional information.

6.5.3   *Coordinate transformations and imbeddings from another point of view*

Let us address the problem of imbedding coordinates going in the "opposite direction". Consider the following formal procedure. From a given dynamical system:

$$\dot{x} = f(x, y, z)$$
$$\dot{y} = g(x, y, z)$$
$$\dot{z} = h(x, y, z),$$

pick one coordinate, e.g., the y-coordinate, and build the following system:

$$\dot{y} = v$$
$$\dot{v} = w \tag{6.2}$$
$$\dot{w} = F(y, v, w),$$

where $v = g(x, y, z)$, $w = \dot{g} = f\frac{\partial g}{\partial x} + g\frac{\partial g}{\partial y} + h\frac{\partial g}{\partial z}$ and $F = \ddot{g}$, in the way it was advanced in the previous Chapter. Formally, we may always hope to solve these three last conditions in terms of the original coordinates. If the procedure can be carried out in such a way that the right-hand side in Eq. (6.2) consists of non-singular, invertible, smooth functions, then we can properly regard the new system as equivalent to the original one, under the coordinate transformation $(x, y, z) \mapsto (y, v, w)$. A time-series recording for the y-coordinate would be the same for both systems, and additionally, a derivative imbedding generates the natural coordinates for the second system.

This formal procedure, when possible, can be seen as a way to extend the ideas developed for periodically forced oscillators. The derivative imbedding (and delay imbeddings with small enough delay) generates a coordinate transformation from the original system, and the imbedding dynamics is the same as the original one.

The Chua oscillator [Chua et al. 1986] is one of the few widely studied systems that can be transformed into a derivative set of coordinates with non-singular right-hand side, retaining the same degree of smoothness (C_0) as the original system. Consider

$$\dot{x} = \alpha(y - h(x))$$
$$\dot{y} = x - y + z \qquad (6.3)$$
$$\dot{z} = -\beta y$$

where $h(x) = \gamma x + \delta(|x + 1| - |x - 1|)$. The set of parameters most widely studied corresponds to $\alpha = 7$, $\gamma = 2/7$, $\delta = -3/14$ and $\beta \in [6.5, 10.5]$. From the third equation we get $y = -\dot{z}/\beta$ and derivating this equation together with the second equation we get $x = -\ddot{z}/\beta - \dot{z}/\beta - z$. Also we have that $\dddot{z} = -\beta\ddot{y} = -\beta(\dot{x} - \dot{y} + \dot{z}) = -\beta(\alpha(y - h(x)) - \dot{y} + \dot{z})$. This last equation can be completely rewritten in terms of z, $\dot{z}$ and $\ddot{z}$, and hence the coordinates $(z, w = \dot{z}, v = \ddot{z})$ can be used to describe the original problem. Solutions of this system can be translated to the original one by computing x and y as prescribed above. However, Chua's oscillator, as in general piecewise-linear systems, is non trivial. The apparent simplicity of the linear portions conceals all the nonlinearities and difficulties lying in the connecting points. This system has in fact periodic orbits that can be read as braids having both positive and negative crossings. A surface of the form *third coordinate equals constant* does not function as a good Poincaré section for this system, and hence, the system lies outside the validity range of e.g., McRobie's Theorem.

The inverse question we posed at the beginning of this Subsection reads now what is the relation between the original system and Eq. (6.2) when its right-hand side does not consist of non-singular, invertible, smooth functions, or otherwise when its Poincaré section is not of the form *third coordinate equals constant?*

6.5.4   *Symmetries*

Chua's oscillator as well as Lorenz equations and Duffing oscillator present a reflexion symmetry, $G = \{\Pi, \Pi^2 = Id\}$. The transformation $(x, y, z) \mapsto -(x, y, z)$ in Chua's oscillator and the change $(x, y, z) \mapsto (-x, -y, z)$ in Lorenz transform solutions of the problem into solutions of the problem. The procedure (6.2) performed on Chua's equations returns a new presentation of the problem with the same symmetry. However, when performed on Lorenz equations, it can return a system with symmetry $(x, y, z) \mapsto -(x, y, z)$ or a system without symmetry. The second case happens if we pick as our first coordinate the variable z while the first case corresponds to the choice of x or y as the first coordinate.

The discussion for delay coordinates shows the same property. If the z coordinate in the Lorenz system is sampled, the delay imbedding will show no symmetry. However, if the x or y coordinate is picked, the symmetry will act on the imbedded system as the symmetry in Chua's oscillator, $(x, y, z) \mapsto -(x, y, z)$.

While Chua's oscillator has been shown to be associated to a universal template [Ghrist and Holmes 1996], the Lorenz attractor has been associated to a template with only positive braids [Ghrist et al. 1997]. In both cases, the template construction is strongly based on the symmetry of the problem. Can the imbedding based in the x coordinate of the Lorenz system be associated to a universal template?

In the Lorenz system the set $(0, 0, z)$ must contain complete orbits as a consequence of the symmetry. The orbits are actually $(0, 0, 0)$ and each half of the z-axis. In the delay imbedding based on the coordinate x (or y) these three orbits are mapped on $(0, 0, 0)$, being the transformation singular. In the delay imbedding based on the coordinate z the map is 2-to-1 for most points since $(x, y, z)(t) \mapsto (z(t), z(t + \tau), z(t + 2\tau))$ and in the same form $(-x, -y, z)(t) \mapsto (z(t), z(t + \tau), z(t + 2\tau))$, i.e., to the same point, the exception being the z-axis. This symmetry argument shows that the Proposition 6.1 stated above cannot be correct since the maps induced by the imbedding procedure are not one-to-one and they can produce important alterations in the topology of the phase space. However, since data is normally collected from attractors, it often happens that the singularities of the transformation are not immediately displayed. In general, properties in the original system that are not picked-up by the collection of data, will still be absent after imbedding this data. Data manipulation cannot reconstruct features that are undetected by the data.

Most of this analysis has been based on [Letellier and Gilmore 2000, Letellier and Aguirre 2002].

6.5.5 *Concluding remarks on the imbedding problem*

The conclusion for this Section is that except for the special case of dynamical systems that admit a derivative imbedding as a proper change of coordinates, and even in that case, only for delay imbeddings having sufficiently small delay parameter, the usual imbedding procedures might produce some degree of interference with the original dynamics. Imbedded time-series generate topological properties of an (hypothetical) imbedded dynamical system, which **may not be** just the original system recast in a new set of coordinates. That is why topologically inequivalent imbeddings of the same data set seem to appear.

Imbeddings cannot be used as black-box tools to analyze scalar data sets of whatever origin. On the contrary, the topological information given by analyzing imbedded data has to be combined with additional knowledge about the system (when available) in order to establish whether this information stems from the original system only. The topological aspects of time-delay imbeddings have not received so far all the attention that they deserve.

6.6 Higher Dimensions: What is Possible?

Linking and braid properties are intrinsically 3-dimensional. In fact, whatever properties that stem from knots or their generalization will not "upgrade" to higher dimensions, since all knots are trivial in dimension four or higher. The question is: What is left in higher dimensions?

Let us start this discussion from the trivial situation. Given a perfectly manageable Poincaré map in $\mathbb{R}^2$ we may just "dress" each point in the Poincaré section with a line, i.e., by making a tensor product with $\mathbb{R}^1$.

The dynamics of these artificially invented lines in $\mathbb{R}^3$ will be an exact copy of the original dynamics of the points in the (original, 2-D) Poincaré section, including braids, links and all the previous material. Nothing has changed, only that we have devised a (perhaps unnecessarily) complicated way of describing points in the Poincaré section as if they were infinitely long straight lines of a higher-dimensional space. There are other ways of devising objects in higher dimensions that could admit a treatment in terms of 3-D knots and braids. However, the important thing is that such objects should be dynamically relevant, such as flow-invariant torus in some 4-D Hamiltonian systems [Ghrist and Young 1998]. We will analyze the dynamical relevance of some constructions below.

The following degree of complication is when this tensor product has a dynamical motivation (i.e., when we do have a natural way of producing dynamical objects of this kind in our problem), as in the situation described in Chapter 2 about highly dissipative systems (the dissipation is represented by a parameter ϵ in Eqs. (3.2)). The higher-dimensional dynamics eventually decays to a Center Manifold and therefore its representation in terms of a tensor product is proper. Braids and braid properties can be recovered without alterations.

This situation is not far-fetched. In fact, most real-life dynamical systems are very dissipative (if you "pull out the plug" the system will eventually stop). Apart from some exceptional situations, there will in general be one direction that is the "less dissipative" (with slowest decay) while the others will decay faster. In such a case, waiting sufficiently long time, we will have a dynamics where all decaying directions except one have already relaxed exponentially close to the Center Manifold, and the dynamical description with or without the $m \geq 1$ decayed directions will be essentially equivalent.

Of course, not all systems behave in the way indicated above. In fact, the concept of *hyperchaos* [Eiswirth et al. 1992] has been proposed to describe systems where the number the slow-relaxing dimensions is larger than one. The next degree of complication appears then when the reduction to the 3-D Center Manifold is not possible. Still, we may use the methods of Chapter 5 or of this Chapter and imbed our data in d dimensions, but now there is no way of making $d < 4$ (or $d < 3$ in the case of the Poincaré map). We have to live with that. To fix ideas, let us think of a reconstructed flow imbedded in four dimensions.

The first methodological problem is that regardless of the imbedding dimension, the dynamical object that is easy to identify, to approximate and to model still remains a one-dimensional manifold, i.e., the periodic orbits of the flow. To "invent" a 2-dimensional invariant manifold to be inferred from the data will always carry with it some degree of arbitrariness and errors to be added on top of the uncertainty with which periodic orbits were obtained from data. There is no natural such object arising "spontaneously" in the dynamics.

Knot, link and braid properties are hence ruled out from the beginning. If we have a very good data sampling, covering phase space in a satisfactory way, we may attempt to model e.g., the stable manifold of our periodic orbit(s), at least locally in a region sufficiently close to the orbit. In this way, we may produce a 2-dimensional strip representing the orbit and its local stable manifold.

6.6.1   *Local torsion*

Perhaps the simplest attempt at extending braids properties to higher dimensions consists in considering a system in $\mathbb{R}^4$, and study the possible generalization of the notion of local torsion in $\mathbb{R}^3$. This is, we want to know in how many different forms the stable and unstable manifold of a saddle in $\mathbb{R}^n$ $(n = 3, 4)$ can wind around after having evolved along one period of the periodic orbit.

When $n = 3$ there are three local-manifolds, each one of dimension one: the centre manifold constituted by the orbit itself, the stable manifold and the unstable manifold of the saddle periodic orbit. Moving along the orbit, the relative orientation of these manifolds may change. Actually, we can consider the subspaces associated to the linearization of the flow, spanned in this case by just three vectors. Hence, moving along the orbit the three basis vectors move according to elements of $SO(3)$, the rotation group in 3 dimensions. Furthermore, it is intuitive to perform this rotation in two steps. The first step changes the orientation of the velocity to its new direction. In the second step, a rotation is performed along the velocity vector. In such a way, $SO(3)$ is described in terms of elements of $SO(3)/SO(2)$ and $SO(2)$. The *torsion* is precisely the (accumulated) rotation associated to the second step. Since after a full turn of the orbit, all the manifolds must be in coincidence with the initial direction, we expect the torsion to be $n2\Pi$ for some integer n. If the eigenvalues associated to the stable and unstable manifold are both positive (this is called a *regular saddle*), the rotation consists of a full loop in $SO(2)$ and since the fundamental group $\pi_1(SO(2)) = Z$ we obtain the homotopically different torsions for a regular saddle to be Z. In case the saddle is of flip type, i.e., it has negative eigenvalues, we can make two turns to the orbit and apply the same reasoning, in this case, the rotation per turn may be a half integer number.

For $n = 4$ the same construction can be performed, with one essential difference, one of the manifolds (stable or unstable) is bi-dimensional. Then, the rotation of the frame can be traced in $SO(4)$. Again, building the rotation as a composition of two steps, one aligning the initial velocity vector to the new direction and the second step rotating the remaining three basis vectors to match the present orientation, the description is given by $SO(4)/SO(3) \times SO(3)$ and it is the latter group, $SO(3)$, the one containing the torsion.

Rotations within the 2-D stable or unstable manifold are of no interest as the system is not forced to come back to the same orientation within the 2-D manifold. We have then to identify rotations in the 2-D manifold. Hence, we end up inspecting $SO(3)/SO(2)$, instead of just $SO(3)$. As in the three dimensional case, we want to classify the ways in which the local manifolds wrap around the orbit, allowing for deformations, i.e., we are interested in the fundamental group $\pi(SO(3)/SO(2))$.

The fundamental group $\pi(SO(3)/SO(2))$ has just two elements: $\{e, a\}$ (where $a^2 = e$). This is a consequence of the exact sequence of a fibration [Rotman 1988], that relates the homotopy groups of the space, fiber and base of a fibration (in this case a quotient space). Hence, according to whether the eigenvalue of the 1-D manifold (stable or unstable) is positive or negative and to whether the loop is homotopic to the identity (e) or to the second element (a), we have only four types of organizations of saddles, according to the choices: regular or flip, "times" e or a. Studies performed using homology theory indicate that the choice e or a correspond to tori or Klein bottles respectively [Mindlin and Solari 1997]. These later studies were performed on simulations of a regular saddle. The flip saddle was not available in the simulations.

6.6.2   *Topological entropy*

Resigning braids we resign a great deal of what we have called *orbit organization*. Links and braids are trivialized and we are left with little more than orbit counting. In terms of computing complexity estimates such as the topological entropy, orbit counting may be good enough.

The concept of Markov partition in phase-space, where one may assign symbolic labels to different regions of phase-space and represent orbits by itineraries is of course not restricted to 3-D dynamical systems. The concept of topological entropy and its relation with box-count dimension (e.g., some limit growth rate of the number of open sets of size ϵ in a covering of the space when $\epsilon \to 0$) still persists in higher dimensions and it can be used to estimate the complexity of a dynamical system in one way or the other. If not the exact value of the topological entropy, at least methods to compute upper bounds of the topological entropy in terms of finite partitions in phase-space [Froyland et al. 2001] or of homology groups [Manning 1975] (see the next Subsection) have been published.

6.6.3 *Homology groups*

Still, we may want to understand the topological properties of the set of periodic orbits hidden in our data. We need some "braidless" method (in the sense that knots "dissolve" into trivial objects in higher dimensions) and one method that appears to jump at hand is to consider the homology groups associated to our data [Muldoon et al. 1993, Sciamarella and Mindlin 1999; 2001].

Given a set of data points in a 3-D Poincaré section, we may recast our data points as 0-cells, they will always be our fundamental object. Joining the 0-cells pairwise by straight lines (edges between points) we produce 1-cells and assembling these in triangles we may identify 2-cells. The whole data is thus regarded as a topological complex. Further information is given by the Poincaré map since it maps the set of 0-cells onto itself.

The computation of the homology groups associated to this complex has been implemented in [Muldoon et al. 1993, Sciamarella and Mindlin 1999; 2001]. This approach gives a different topological characterization of the data set. It is not only braidless but even "periodic orbit"-less, since the data is characterized as a whole, regardless of the existence of hidden periodic orbits in it. Another advantage is that the procedure is not restricted to 2-D Poincaré maps, it can be applied in any dimension. The "problem", if we want to call it this way, is that the characterization is *different* from that obtained in Chapters 2, 3 and 4. We obtain other information. To translate these new information in the "old one" may be difficult (or irrelevant), hence the method requires to develop an intuition of its own. On the other hand, for the specific goal of validating models and characterizing data, it may be equally good.

The computation of Trees, Fat Trees and their images by a reference 2-D homeomorphism as in Chapter 4 may be regarded as analyzing the consequences of mapping 0-cells by (a representative of) the Poincaré map along with the consequences that this mapping operates on the 1-cells. In the end, we are just considering vertices, edges and their images *recasted as vertices and edges*. However, it is not known to the authors whether there is a deeper connection or not. Some progress in this direction has been advanced in [Lefranc 2006].

Chapter 7

The User's Chapter

The distinction between users and producers of methods, for the analysis of dynamical systems, is to some extent arbitrary. The idea that methods can be taken off-the-shelf and applied to a problem, the black-box dream, has repeatedly emerged in nonlinear dynamics just to be proven wrong as many times as it has arisen.

We want to devote this Chapter to the work performed in analyzing experimental data in terms of the methods exposed in the previous Chapters. We will also present a few other uses of the same ideas.

The orbit reconstruction and topological characterization programme was proposed in a series of papers [Solari and Gilmore 1988b;a, Mindlin et al. 1990]. The first two using experimental data were [Mindlin et al. 1991] studying the Belousov-Zhabotinskii experiment and [Tufilaro et al. 1991] studying the NMR-laser.

Because of social reasons, most of the earliest applications were performed in laser physics [Tufilaro et al. 1991, Papoff et al. 1992, Lefranc et al. 1994, Boulant et al. 1997a;b, Gilmore et al. 1997, Mendez et al. 2001, Amon and Lefranc 2004] but soon the proposed methods began to migrate to other fields of applications such as: astrophysics [Boyd et al. 1994], biology [Trevisan et al. 2005], chemistry [Letellier et al. 1995, Firle et al. 1996, Deshmukh et al. 2001], plasma physics [Letellier et al. 2001] and even economics [Gilmore 2001]. Other works addressed the key problem of the relations with imbeddings [Letellier et al. 1998, Krmpotic and Mindlin 1997, Letellier and Gouesbet 1996, Letellier et al. 1996] or attempted alternative procedures towards the same ends exposed in this book [Gouesbet and Letellier 1994, Sciamarella and Mindlin 1999, McRobie 1992, Carroll 1998, Sciamarella and Mindlin 2001].

In what follows we produce some pointers to the experimental physics literature. The discussion is not intended to be a re-presentation of these works but just a guide to them. The contributions of the theoretical papers has been taken into account in Chapters 5 and 6.

7.1   Laser Physics

The earliest works on relative rotation rates were generated in the quest towards understanding a model for the laser with modulated losses [Solari and Gilmore 1988b] and it was this connection which sparked the interest in the laser physics community.

Lasers with low Fresnel-number are well known for being described in terms of a few variables by the rate equations. Due to the different characteristic decay times of (a) the electric field, (b) the atomic polarization and (c) the excess of molecules in the excited state, the dynamics of lasers was early classified into three classes [Tredicce et al. 1985]: Class A described by the evolution of the light intensity, Class B that requires the inclusion of the number of excited molecules and Class C that requires also the inclusion of the electrical polarization as an independent dynamical variable. Lasers of class A and B are described by sets of ODEs of dimension lower than 3 and display chaos when perturbed, making them an excellent field to attempt topological characterizations.

Papoff et al. [Papoff et al. 1992] studied the laser with saturable absorber, in what appears to be the first application in laser physics. The intensity of the laser beam was sampled at regular intervals, $\{y_i\}$, and orbits reconstructed with the imbedding

$$X_i = y_i$$
$$Y_i = y_i - y_{i-1}$$
$$Z_i = \sum_{j=1}^{i} \exp(-(i-j)/N)y_i,$$

an imbedding introduced in [Mindlin et al. 1991]. The data admitted a global Poincaré section and the flow was associated with a Smale horseshoe template.

A simple extension of the argument of section 6.5.1.1 shows that if a Poincaré section $Z = c$ exists, the resulting braids would be positive (an argument not known at that time). Hence, the hardest part of the task was actually performed when a differential imbedding was found, remaining only the task of identifying the number of leaves, the torsion and gluing of the template leaves.

The CO_2 laser with modulated losses was considered in [Lefranc et al. 1994]. The laser is in class B, the variables are then the light intensity and the population inversion, while the third dimension corresponds to the periodic modulation of the losses by an electro-optical modulator inserted in the cavity. When the frequency of the modulation is comparable with the relaxation frequency of the laser, chaos had been observed and described [Arecchi et al. 1982] and the corresponding equations studied [Solari et al. 1987] previously. Furthermore, relative rotation rates have been introduced using this laser as an example [Solari and Gilmore 1988b] and the model was associated to a horseshoe template. We can say then that the theory was in agreement with the experiment. The kneading frequencies of the horseshoe were used to characterize a sudden change of the attractor, called a *crisis* [Grebogi et al. 1982], involving period two and three orbits. The crisis was also predicted by the theory.

Lefranc et al. [Lefranc et al. 1994] studied the chaotic attractor at different parameter values finding that in all the cases the horseshoe signature was present. Additionally, they were able to identify an (approximated) generating partition of the phase space as well as to identify the transformations suffered by the chaotic attractor in several crisis. The signature of the attractor was described in terms of symbolic sequences pertaining to Smale's horseshoe.

Boulant et al. [Boulant et al. 1997b] considered data from Nd-doped fiber laser with pump modulation at different frequencies, roughly corresponding to $1/4$, $1/3$ and $1/2$ the relaxation frequency, w_r, of the laser. In the three cases chaotic attractors were found with an organization compatible with a horseshoe but with different global torsions Θ_g, related to the order of the sub-harmonic in the form $\Theta_g(C_{1/n}) = n - 1$ where $C_{1/n}$ is the attractor associated to the forcing of frequency w_r/n.

Boulant et al. [Boulant et al. 1997a] also considered a $Nd : YAG$ laser with pump modulation. The imbedding proposed was of the type $X, \dot{X}, \phi$ with X the light intensity and ϕ the phase of the modulation. The imbedding was assumed to be a good one but no checks were performed. Note however that *two experimental variables were used* (this effort of collecting all available experimental information is a good practice that partially relieves the difficulties of the imbedding procedure). The braids were then necessarily positive and the template was identified as having two branches, one preserving order with a full turn of torsion and the other reversing order with half a turn of torsion, with the order preserving branch glued behind the reversing branch. The authors identified a knotted period-3 orbit along with unknotted orbits (e.g., period-1 orbits). The braid associated to the knotted orbit can be identified as $(\sigma_1\sigma_2)^2$, with associated Conway polynomial $1 + z^1$. This work shows, as novelty, a template with positive braids which is not the simplest horseshoe template.

A triply resonant optical parametric oscillator (TROPO) was studied in [Amon and Lefranc 2004]. Long time before the experiment, theoretical models predicted the existence of chaos in this optical device. However, no experimental confirmation had been produced before. One of the main difficulties was that the system experiences parameter drifts on a time-scale comparable to the mean dynamical period, hence the stationarity of the time-series, usually required in most methods of time-series analysis, was not present. The authors noticed however that the drift was sufficiently slow as to permit a reconstruction for relatively short periods of time, and even in such small data-sets, it was still possible to extract periodic orbit representatives. The orbits were reconstructed in the standard time-delay imbedding and a horseshoe-like organization was found. More interesting, some of the braids were "chaotic" in the sense that they imply a positive topological entropy. Hence, the presence of chaos was confirmed through the use of topological methods.

7.2   Other Experiments

7.2.1   *Biological application*

Within the vast field of biology, there exist some applications where topological data analysis seems to be relevant. In [Trevisan et al. 2005] the goal of the authors is to identify speakers by extracting some "signature"-like characteristic from their utterances. Voice data (from human speech) is recasted as a data-series via $x(f) = \ln|H(f)|^2$, where $H(f)$ is the Fourier transform of the original recorded data-set, and f, the frequency, plays the role of independent variable. A delay imbedding of this set allowed the authors to identify periodic orbits of low period and their linking properties (relative rotation rates and linking numbers). They assigned to each speaker an array of numbers, listing the linking properties found through the analysis of vowel utterances. This is a surprising application, since,

---

[1]Improperly identified as the trefoil knot type, which has the polynomial $1 + z^2$.

a priori, there were no reasons to believe that voice data would present a "low-dimensional" dynamical behaviour.

7.2.2 Chemical data

Copper electrodisolution was studied in [Letellier et al. 1995]. In this case a horseshoe template was identified and the orbit organization used as a test to validate a model obtained by fitting a polynomial vector field to the same problem. The original proposal of using orbit organization analysis to partially validate models [Solari and Gilmore 1988b] was acted for the first time.

The catalytical reaction of CO and O_2 on a $Pt(110)$ surface was studied in [Firle et al. 1996]. The imbedding consisted of the measured data point, x_1, a convolution of the data with an exponential damping factor, x_2, and a Hilbert transform, x_3. The braids identified not only presented positive crossings, as we had become familiar with at that time, but alternating both negative and positive crossings. The Poincaré section was easy to identify in the (x_1, x_2)-plane. The period three orbit corresponded to the braid $\sigma_1\sigma_2^{-1}$ implying positive topological entropy.

In [Deshmukh et al. 2001] stress dynamics of polymer solutions was considered, reconstructing the phase space with the data, x_1, an integral of the data, x_2 and the derivative. For this imbedding Theorem 6.1 is applicable in principle. The template identified corresponds to a horseshoe.

7.2.3 Plasma physics

A thermionic diode plasma experiment has been investigated in terms of a branched manifold schemed by a template [Letellier et al. 2001]. The imbedding was of differential type $(x, \dot{x}, \ddot{x})$, having hence only positive braids. The orbits were found to be organized in a three-branched template, i.e., a structure more complicated than the common horseshoe template.

Chapter 8

After Thoughts

If we were to advise a newcomer willing to analyze data with the methods discussed in this book, what are the guidelines we should emphasize?

- Have a talk with the experimentalist supervising the data collection and collect all available data. Do not stop at the time-series for $x(t)$ if you can record two or all three coordinates.
- Have a talk with other theoreticians and gather all general information about your system. Be suspicious and criticize. We theoreticians have a weakness for deceiving ourselves, which you should **not** imitate. Test this general information against your collected data, keep only the compatible subset of theoretical information. Also, try to understand what was wrong with the incompatible subset; at least the authors of this book will show you their gratitude.
- Go through the methods of Chapters 5, 6 and 2 to 4, keeping only the results compatible with the previous points.
- Be aware of the fact that still after these precautions, these methods may generate several different inequivalent descriptions of the data (see below for more comments in this direction). Some information may be imbedding independent while some other may not. Remember then that your description(s) do not reflect properties of the data only, but rather of the "data + imbedding" apparatus.

The programme of reconstructing topological aspects of 3-D experimental systems rests on two well-defined columns, the phase space reconstruction and the classification itself.

The topological characterization of diffeormophisms of the disc, or equivalently, of flows with a global Poincaré surface homotopic to a disc was completed by Thurston. However, in more general cases where a global

Poincaré surface does not exist, an equivalent classification using knots is still pending, and it is one area that deserves further attention.

Although the imbedding problem and the reconstruction of orbits in phase space was considered to be a relatively simpler and secondary matter at the beginning of the programme, the results are persuasive: the topological structure carried by experimental data depends on the imbedding, and presumably, not every imbedding can be used for every flow (even after the known incompatibilities have been ruled out). Is it possible to establish equivalence classes among imbeddings? The imbedding problem has emerged with force and it is crucial for the conclusion of the programme.

Other open questions raised by the imbedding procedure could be: (a) Can two "different" dynamical systems generate the same dynamics in one of the coordinates? In other words, does there exist a pair of dynamical systems in 3-D, described by the coordinates x, y, z and x, v, w respectively, such that (i) v, w differ from x, y in more than a trivial coordinate transformation (whatever "more" could mean in this context) and (ii) the x-component of the orbit of the first system going through (x_0, y_0, z_0) coincides with the orbit of the second system going through (x_0, v_0, w_0)? Posed in this broad form, the answer is "yes", we have discussed some ideas in this direction in Chapter 6. How can these ideas be developed in order to gain understanding about the relationship between a time-series, its originating dynamical system and the data-collecting function? (b) The topological analysis is most profitable when the system has a global Poincaré section isomorphic to a disc. Then the analysis goes along the $\mathbb{R}^2 \times \mathbb{S}^1$ problem. However, this is not always the case, and global Poincaré sections can be of other types. How to proceed in these cases? how do we induce from the collected data the characteristics of the proposed Poincaré section?

On the other hand, templates have been useful tools but they are not the answer they once appeared to be. The existence of infinitely many universal templates is an obstacle. It forces to consider these templates as competing physical theories, but quite often there is no mechanism to decide among these competing theories, since their predictions differ exactly where there is no available data.

Another aspect that deserves further attention is that through the method of using only (reconstructed) periodic orbits, much of the data available is discarded (the portions of the data outside the close returns). Ways of including in the topological analysis this additional data is badly needed.

In short, as we advanced in the preface, the programme is alive and still incomplete.

Bibliography

Abarbanel, H. D. I., Brown, R., Sidorowich, J. J., Tsimring, L. S., 1993. The analysis of observed chaotic data in physical systems. Rev. Mod. Phys. 65, 1331.

Adams, C. C., 2001. The Knot Book. W H Freeman, New York.

Amon, A., Lefranc, M., 2004. Topological signature of deterministic chaos in short nonstationary signals from an optical parametric oscillator. Physical Review Letters 92 (9), 094101.

Arecchi, F. T., 1988. Instability and chaos in optics. Physica Scripta T23, 160–164.

Arecchi, F. T., Meucci, R., Gadomski, W., 1986. Laser dynamics with competing instabilities. Phys. Rev. Lett. 58 (21), 2205–2208, laser, Shilnikov.

Arecchi, F. T., Meucci, R., Puccioni, G., Tredicce, J., 1982. Experimental evidence of subharmonic bifurcations, multistability, and turbulence in a q-switched gas laser. Phys. Rev. Lett. 49, 1217.

Aubry, N., Guyonnet, R., Lima, R., 1991. Spatiotemporal analysis of complex signals: Theory and applications. Journal of Statistical Physics 64 (3-4), 683–739.

Baldwin, G. C., 1969. An Introduction to Nonlinear Optics. Plenum, New York.

Bestvina, M., Handel, M., 1992. Train tracks and automorphisms of free groups. Annals of Mathematics 135, 1–51.

Bestvina, M., Handel, M., 1995. Train tracks for surface homeomorphisms. Topology 34, 109–140.

Birman, J. S., Williams, R. F., 1983a. Knotted periodic orbits in dynamical systems i: Lorenz equations. Topology 22 (1), 47–82.

Birman, J. S., Williams, R. F., 1983b. Knotted periodic orbits in dynamical systems ii: knot holders for fibered knots. Contempory Mathematics 20,

1–60.

Block, L., Guckenheimer, J., Misiurewicz, M., Young, L. S., 1980. Periodic points and topological entropy of one dimensional maps. Lecture Notes Math. 819, 18.

Boulant, G., Bielawski, S., Derozier, D., Lefranc, M., 1997a. Experimental observation of a chaotic attractor with a reverse horseshoe topological structure. Physical Review E 55 (4), R3801–3804.

Boulant, G., Lefranc, M., Bielawski, S., Derozier, D., 1997b. Horseshoe templates with global torsion in a driven laser. Physical Review E 55 (5), 5082–5091.

Boyd, P., Mindlin, G. B., Gilmore, R., Solari, H. G., 1994. Topological analysis of chaotic orbits: Revisiting hyperion. The Astrophysical Journal 431, 425–431.

Boyland, P., 1984. Braid types and a topological method of proving positive topological entropy, preprint, Department of Mathematics, Boston University.

Carlson, S. C. (Ed.), 2001. Topology of Surfaces, Knots and Manifolds. J Wiley and Sons, New York.

Carroll, T. L., 1998. Approximating chaotic time series through unstable periodic orbits. Physical Review E 59, 1615–1621.

Casson, A., Bleiler, S., 1988. Automorphisms of Surfaces after Nielsen and Thurston. Cambridge University Press, Cambridge.

Chua, L. O., Komuro, M., Matsumoto, T., 1986. The double scroll family. IEEE Transactions on Circuits and Systems CAS-33 (11), 1073–1118.

Collet, P., Eckman, J.-P., 1986. Iterated Maps of the Interval as Dynamical Systems. Birkhäuser, Basel.

de Carvalho, A., Hall, T., 2003. Conjugacies between horseshoe braids. Nonlinearity 16, 1329–1338.

de Carvallo, A., Hall, T., 2001. Pruning theory and thurston's classification of surface homeomorphisms. J. Eur. Math. Soc. 3, 287–333.

de Carvallo, A., Hall, T., 2002. How to prune a horseshoe. Nonlinearity 15, R19–R68.

Deshmukh, S., Ghosh, A., Badiger, M. V., Kumar, V. R., Kulkarni, B. D., 2001. Characterization ofchaotic dynamics ii: topological invariants and their equivalence for an autocatalytic model system and an experimental sheared polymer solution. Chemical Engineering Systems 56, 5643–5651.

Eiswirth, M., Kr̈uel, T.-M., Ertl, G., Schneider, F.-W., 1992. Hyperchaos in chemical reactions. Chemical Physical Letters 193, 305.

Fathi, A., Shub, M., 1979. Some dynamics of pseudo-anosov diffeomorphisms. Asterisque 66-67, 181–207.

Feigenbaum, M. J., 1978. Quantitative universality for a class of nonlinear transformations. J. Stat. Phys. 19, 25.

Fioretti, A., Molesti, F., Zambon, B., Arimondo, E., Papoff, F., 1993. Topological analysis of laser with saturable absorber in experiments and models. Int. J. Bifurcation Chaos Appl. Sci. Eng. 3, 559–564.

Firle, S. O., Natiello, M. A., Eiswirth, M., 1996. Topological dynamics in a catalysys experiment. Phys. Rev. E53, 1257.

Franks, J., Misiurewicz, M., 1993. Cycles for disk homeomorphisms and thick trees. Contemporary Mathematics 152, 69–139.

Froyland, G., Junge, O., Ochs, G., 2001. Rigorous computation of topological entropy with respect to a finite partition. Physica D 154, 68–84.

Gambaudo, J. M., van Strien, S., Tresser, C., 1989. The periodic orbit structure of orientation preserving diffeomorphisms on D^2 with topological entropy zero. Ann. Inst. Henri Poincaré Phys. Théor. 49, 335.

Ghrist, R., Holmes, P., 1996. An ode whose solutions contain all knots and links. International Journal of Bifurcation and Chaos 6 (5), 779–800.

Ghrist, R., Young, T., 1998. From morse-smale to all knots and links. Nonlinearity 11, 1111–1125.

Ghrist, R. W., Holmes, P. J., Sullivan, M. C., 1997. Knots and Links in Three-dimensional flows. Vol. 1654 of Lecture Notes in Mathematics. Springer Verlag, Berlin.

Gilmore, C. G., 2001. An examination of nonlinear dependence in exchange rates, using recent methods from chaos theory. Global Finance Journal 12, 139–151.

Gilmore, R., Vilaseca, R., Corbalan, R., Roldan, E., 1997. Topological analysis of chaos in the optically pumped laser. Phys. Rev. 55 E, 2479–2487.

Gouesbet, G., Letellier, C., 1994. Global vector-field reconstruction by using a polymodal l_2 approximation on nets. Physical Review E 49, 4955–4972.

Grebogi, C., Ott, E., York, J. A., 1982. Chaotic attractors in crisis. Physical Review Letters 48 (12), 1507–1510.

Guckenheimer, J., Holmes, P. J., 1986. Nonlinear Oscillators, Dynamical Systems and Bifurcations of Vector Fields. Springer, New York, 1st printing 1983.

Hale, J. K., 1969. Ordinary Differential Equations. Wiley, New York.

Hall, T., 1994a. The creation of horseshoes. Nonlinearity 7, 861.

Hall, T., 1994b. Fat one-dimensional representatives of pseudo-anosov isotopy classes with minimal periodic orbit structure. Nonlinearity 7, 367–

384.

Hirsh, M. W., Smale, S., 1978. Differential Equations, Dynamical Systems, and Linear Algebra. Academic, New York.

Holmes, P., 1989. Knoted periodic orbits in suspensions of smales horseshoe: Extended families and bifurcation sequences. Physica D 40, 42–64.

Holmes, P., Williams, R. F., 1985. Knotted periodic orbits in suspensions of Smale's horseshoe: torus knots and bifurcation sequences. Arch. Rational Mech. Anal. 90, 115.

Katok, A., 1980. Liapunov exponents, entropy and periodic orbits for diffeomorphisms. Publ. Math. IHES 51, 137.

Kauffman, L. H., 1991. Knots and Physics. World Scientific, Singapore.

Kirillov, A. A., 1976. Elements of the Theory of Representations. Springer-Verlag, New York.

Krmpotic, D., Mindlin, G. B., 1997. Truncating expansions in bi-orthogonal bases: What is preserved? Physics Letters A 236, 301–306.

Lakatos, I., 1978. Mathematics, Science and Epistemology. Cambridge University Press, Cambridge, U.K.

Lathrop, D. P., Kostelich, E. J., 1989. Analyzing periodic saddles in experimental strange attractors. In: Abraham, N. B., Albano, A. M., Passamante, A., Rapp, P. E. (Eds.), Measures of Complexity and Chaos. Vol. B 208 of NATO ASI Series. Plenum, New York, p. 147.

Lefranc, M., 2006. Alternative determinism principle for topological analysis of chaos. Phys. Rev. E 74, 035202.

Lefranc, M., Glorieux, P., Papoff, F., Molesti, F., Arimondo, E., 1994. Combining topological analysis and symbolic dynamics to describe a strange attractor and its crises. Phys. Rev. Lett. 73, 1364.

Letellier, C., Aguirre, L. A., 2002. Investigating nonlinear dynamics from time series: The influence of symmetries and the choice of observables. Chaos 12 (3), 549–558.

Letellier, C., Dutertre, P., Reizner, J., Gouesbet, G., 1996. Evolution of a multimodal map induced by an equivariant vector field. J. Phys. A: Math. Gen. 29, 5359–5373.

Letellier, C., Gilmore, R., 2000. Covering dynamical systems: Twofold covers. Physical Review E 63, 016206.

Letellier, C., Gouesbet, G., 1996. Topological characterization of reconstructed attractors modding out symmetries. J. Phys-II France 6, 1615–1638.

Letellier, C., Maquet, J., Sceller, L. L., Gouesbet, G., Aguirre, L. A., 1998. On the non-equivalence of observables in phase-space reconstructions from recorded time series. Journal of Physics A; Math. Gen. 31, 7913–7927.

Letellier, C., Mnarda, O., Klinger, T., Piel, A., Bonhommed, G., 2001. Dynamical analysis and map modeling of a thermionic diode plasma experiment. Physica D 156, 169–178.

Letellier, C., Sceller, L. L., Marchal, E., Dutertre, P., Gouesbet, G., Fei, Z., Hudson, J. L., 1995. Global vector-field reconstruction from a chaotic experimental signal in copper electrodissolution. Physical Review E 51, 4262–4266.

Li, T.-Y., Yorke, J., 1975. Period three implies chaos. Amer. Math. Monthly 82, 985.

Lorenz, N., 1963. Deterministic non-periodic flow. J. Atmospheric Science 20, 130–141.

Los, J. E., 1993. Psudo-anosov maps and invariant train tracks in disks: a finite algorithm. Proc. London Math. Soc. 66, 400–430.

Manning, A., 1975. Topological entropy and the first homology group. In: Dold, A., Eckmann, B. (Eds.), Dynamical Systems - Warwick 1974. Springer Verlag, pp. 185–190, lecture Notes in Mathematics, 468.

McRobie, F. A., 1992. Bifurcation precedences in the braids of periodic orbits of spiral 3-shoes in driven oscillators. Proceedings: Mathematical and Physical Series 438, 545–569.

McRobie, F. A., Thompson, J. M. T., 1993. Braids and knots in driven oscillators. International Journal of Bifurcation and Chaos 3(6), 1343–1461.

Mendez, J. M., Laje, R., Giudici, M., Aliaga, J., Mindlin, G. B., 2001. Dynamics of periodically forced semiconductor laser with optical feedback. Physical Review E 63, 066218.

Metropolis, N., Stein, M. L., Stein, P. R., 1973. On finite limit sets for transformations on the unit interval. J. Combinatorial Theory 15A, 25.

Mindlin, B. G., Solari, H. G., 1997. Tori and klein bottles in four dimensional chaotic flows. Physica D 102, 177.

Mindlin, G., Solari, H., 1995. Topologically inequivalent embeddings. Phys. Rev. E52, 1497.

Mindlin, G. B., Hou, X.-J., Gilmore, R., Solari, H. G., Tufillaro, N., 14 May 1990. Classification of strange attractors by integers. Physical Review Letters 64, 20.

Mindlin, G. B., Lopez-Ruiz, R., Solari, H. G., Gilmore, R., 1993. Horseshoe implications. Phys. Rev. 48A, 4297.

Mindlin, G. B., Solari, H. G., Natiello, M., Gilmore, R., Hou, X.-J., 22 April

1991. Topological analysis of chaotic time series data from the belusov-zhabotinskii reaction. Journal of Nonlinear Sciences 1, 147–173.

Misiurewicz, M., 1997. Remarks on sharkovsky's theorem, published in American Mathematical Monthly, Nov 1997.

Muldoon, M. R., MacKay, R. S., Broomhead, D. S., Huke, J. P., 1993. Topology from time series. Physica D 65, 1–16.

Narayanan, K., Govindan, R., Gopinathan, M., 1998. Unstable periodic orbits in human cardiac rhythms. Phys. Rev. E 57 (4), 4594–4602.

Natiello, M. A., Solari, H. G., 1994. Remarks on braid theory and the characterisation of periodic orbits. J. Knot Theory Ramifications 3, 511.

Oppo, G.-L., Politi, A., 1989. Center-manifold reduction for laser equation with detuning. Physical Review A 40, 1422–1427.

Oppo, G.-L., Politi, A., Lippi, G.-L., Arecchi, F. T., 1986. Frecuency pushing in lasers with injected signal. Physical Reviews A 34, 4000.

Papoff, F., Fioretti, A., Arimondo, E., Solari, G. B. M. H. G., Gilmore, R., 1992. Structure of chaos in the laser with saturable absorber. Phys. Rev. Letters 68, 1128–1131.

Plumecoq, J., Lefranc, M., 2000a. From template analysis to generating partitions i: Periodic orbits, knots and symbolic encodings. Physica D 144, 231–258.

Plumecoq, J., Lefranc, M., 2000b. From template analysis to generating partitions ii: Characterization of the symbolic encodings. Physica D 144, 259–278.

Popper, K., 1959. The Logic of Scientific Discovery. Routledge, London, first edition 1934.

Risken, H., 1989. The Fokker–Planck equation, Methods of Solution and Application (Springer Series in Synergetics, V.18), 2nd edition. Springer, Heidelberg.

Rotman, J. J., 1988. An Introduction to Algebraic Topology. Springer, New York.

Rychlik, M. R., 1989. Lorenz attractors through Šil'nikov-type bifurcation. part I. Ergod. Th. & Dynam. Sys. 10, 793.

Saltzman, B., 1962. Finite amplitude free convection as an initial value problem I. J. Atmos. Sci. 19, 329.

Sauer, T., Yorke, J. A., Casdagli, M., 1991. Embedology. J. Stat. Phys. 65, 579.

Sciamarella, D., Mindlin, G. B., 1999. Topological structure of chaotic flows from human speech data. Phys. Rev. Lett. 82 (7), 1450–1453.

Sciamarella, D., Mindlin, G. B., 2001. Unveiling the topological structure of chaotic flows from human speech data. Phys. Rev. E 64, 036209–1 to 10.

Smale, S., 1963. Diffeomorphisms with many periodic points. In: Cairns, S. S. (Ed.), Differential and Combinatorial Topology. Princeton University, Princeton, p. 63.

Solari, H., Natiello, M., Mindlin, B., 1996a. Nonlinear Dynamics: A Two-way Trip from Physics to Math. Institute of Physics, Bristol.

Solari, H. G., Eschenazi, E., Gilmore, R., Tredicce, J. R., 1987. Influence of coexisting attractors on the dynamics of a laser system. Opt. Commun. 64, 49.

Solari, H. G., Gilmore, R., 1988a. Organization of periodic orbits in the driven duffing oscillator. Phys. Rev. 38A, 1566–1572.

Solari, H. G., Gilmore, R., 1988b. Relative rotation rates for driven dynamical systems. Phys. Rev. 37A, 3096.

Solari, H. G., Natiello, M. A., 2005. Minimal periodic orbit structure of 2-dimensional diffeomorphisms. Journal of Nonlinear Science 15(3), 183–222.

Solari, H. G., Natiello, M. A., Vazquez, M., 1996b. Braids on the poincaré section: A laser example. Phys. Rev. E54, 3185.

Sparrow, C., 1982. The Lorenz Equations: Bifurcations, Chaos, and Strange Attractors. Springer-Verlag, New York.

Takens, F., 1981. Detecting strange attractors in turbulence. In: Rand, D. A., Young, L.-S. (Eds.), Lecture Notes Math. Vol. 898. Springer, Berlin, p. 366.

Tredicce, J. R., Arecchi, F. T., Lippi, G. L., Puccioni, G. P., 1985. Instabilities in lasers with an injected signal. Journal Optical Society of America B 2, 173–183.

Tredicce, J. R., Arecchi, F. T., Puccioni, G. P., Poggi, A., Gadomski, W., 1986. Dynamic behavior and onset of low dimensional chaos in a modulated homogeneously broadened single mode laser: experiments and theory. Phys. Rev. A 34, 2073–2081.

Trevisan, M. A., Eguia, M. C., Mindlin, G. B., 2005. Topological voiceprints for speaker identification. Physica D 200, 75–80.

Tsankov, T. D., Nishtala, A., Gilmore, R., 2005. Embeddings of a strange attractor into r^3. Physical Review E 69, 056215–1:8.

Tucker, W., 1999. The lorenz attractor exists. C. R. Acad. Sci. Paris Ser. I Math 328, 1197–1202.

Tufilaro, N. B., Holzner, R., Flepp, L., Brun, R., Finardi, M., Badii, R., 1991. Template analysis for a chaotic nmr laser. Phys. Rev. 44 A, R4786–

4788.

Šarkovskii, A. N., 1964. Coexistence of cycles of a continuous map of a line into itself. Ukr. Mat. Z. 16, 61.

Whitney, H., 1936. Differentiable manifolds. Ann. Math. 37, 654.

Williams, R. F., 1977. The structure of lorenz attractors. Springe Lecture Notes in Math. In Turbulence Seminar, Berkeley 1976/77 A. Chorin, J.E. Mardsen, S. Smale (eds) 615, 49–116.

Williams, R. F., 1979. The structure of lorenz attractors. Institute Hautes Etudes Sci. Publ. Math. 50, 73–99.

Index